STUDENT'S SOLUTIONS MANUAL
TO ACCOMPANY

MULTIPLE-CHOICE & FREE-RESPONSE QUESTIONS IN PREPARATION FOR THE AP CALCULUS (AB) EXAMINATION

(NINTH EDITION)

By David Lederman

With the assistance of:

Lin McMullin

D&S MARKETING SYSTEMS, INC.
1205 38th Street • Brooklyn, NY 11218

www.dsmarketing.com

ISBN # 1-934780-09-X

Printed in the U.S.A.

PREFACE

This Student's Solution Manual is a supplement to MULTIPLE-CHOICE & FREE-RESPONSE QUESTIONS IN PREPARTION FOR THE AP CALCULUS (AB) EXAMINATION, NINTH EDITION.

This manual is a valuable supplement for those students who desire to see the steps involved in working out the problems. The student should try to become independent of this manual. Only use it after you have tried a problem and are "stuck".

I will be delighted to hear from students and teachers who come up with neat and elegant ways of solving the problems in this workbook. I would also appreciate hearing from you if you find any errors, typographical or mathematical.

I wish to acknowledge several people whose help and encouragement were invaluable in the production of this manual. I wish to thank Bob Byrne of St. Thomas Aquinas HS (Fort Lauderdale, FL) for sharing his ideas and making valuable suggestions for some solutions which have been incorporated in this edition. A special thanks to Lin McMullin for submitting solutions to the questions which he submitted for this edition. I wish to thank Tammi Pruzansky of D&S Marketing Systems, Inc., for preparing the graphs and typesetting the entire manuscript. Finally, I am grateful to my wife for her encouragement and understanding of the many hours required in the production of this manual.

Any errors found in this book are solely the responsibility of the author.

All communications concerning this book should be addressed to:

D & S Marketing Systems, Inc.
1205 38th Street
Brooklyn, NY 11218
www.dsmarketing.com

TABLE OF CONTENTS

Sample Examination I

1. To be continuous both parts of the function must have the same value at $x = 5$.

$$\begin{aligned} 5^2 + 5b &= 5\sin\left(\tfrac{\pi}{2}(5)\right) \\ 25 + 5b &= 5(1) \\ 5b &= -20 \\ b &= -4 \end{aligned}$$

The correct choice is (C).

2. $y = 3x^2 - x^3$

$y' = 6x - 3x^2 = 3x(2 - x)$

The critical points are at $x = 0$ and $x = 2$.

Using the First Derivative Test to find possible relative max/min,

	$x < 0$	$0 < x < 2$	$x > 2$
y'	$-$	$+$	$-$

$x = 2$ is the only relative maximum (where y' switches from + to –).

Therefore, (2,4) is the only relative maximum of $y = 3x^2 - x^3$.

The correct choice is (C).

3. Consider a thin strip of the forest x_i miles from the longer edge. This strip has an area of $5\Delta x$ square miles. The density gives the number of trees per square mile, so the number of trees in this strip is $\rho(x_i)(5\Delta x)$. The sum of all such strips is $\sum_{i=1}^{n}\rho(x_i)(5\Delta x)$. The limit of this Riemann sum is $5\int_0^3 \rho(x)\,dx$ and this gives the total number of trees in the forest.

The correct choice is (C).

4. $f(x) = e^{\sin x}$

   $f'(x) = e^{\sin x} \cos x$

   To find the zeros of $f'(x)$ on the closed interval $[0,2\pi]$, set $f'(x) = 0$.

   $$(e^{\sin x})(\cos x) = 0 \text{ when } \cos x = 0 \text{ or } x = \frac{\pi}{2}, \frac{3\pi}{2} \text{ in the interval } [0,2\pi].$$

   Note: $e^{\sin x}$ is never equal to zero.

   Consequently, $f'(x)$ has two zeros on $[0,2\pi]$.

   The correct choice is (B).

5. As $x \to \infty$, the denominator approaches ∞ faster than the numerator, since the degree of the polynomial in the denominator is greater than the degree of the polynomial in the numerator.

   Therefore, the entire expression approaches 0 as $x \to \infty$.

   The correct choice is (A).

6. The function decreases, then increases and then decreases again. Therefore the derivative should be negative, then positive and then negative again. This is graph (B).

   The correct choice is (B).

7. $f(x) = \sqrt{4 \sin x + 2} = (4 \sin x + 2)^{\frac{1}{2}}$

   $f'(x) = \frac{1}{2}(4 \sin x + 2)^{-\frac{1}{2}}(4 \cos x) = \dfrac{2 \cos x}{\sqrt{4 \sin x + 2}}$

   Therefore, $f'(0) = \dfrac{2 \cos(0)}{\sqrt{4 \sin(0) + 2}} = \dfrac{2(1)}{\sqrt{0 + 2}} = \dfrac{2}{\sqrt{2}}$ or $\sqrt{2}$

   The correct choice is (E).

8. The expression $\pi(R(x))^2$ is the cross section area of the pipe. Since x measures the position at which the cross section is measured, integrating in terms of x from $x = 10{,}000$ to $x = 30{,}000$ gives the volume of that section of the pipe, not necessarily the amount of water flowing through the pipe.

   The correct choice is (B).

9. By implicit differentiation of $x^2 + y^2 = 169$, $2x + 2y\dfrac{dy}{dx} = 0$

   Solve for $\dfrac{dy}{dx}$: $\dfrac{dy}{dx} = \left.\dfrac{-x}{y}\right|_{(5,-12)} = \dfrac{5}{12}$

   The tangent line to the circle $x^2 + y^2 = 169$ at (5, –12) has a slope of $\dfrac{5}{12}$.

   The equation of the tangent line is $y - y_1 = m(x - x_1)$, or

   $$y - (-12) = \frac{5}{12}(x - 5) \text{ or } y + 12 = \frac{5}{12}(x - 5).$$

   Multiply both sides of the equation by 12 and simplify, thus $5x - 12y = 169$.

   The correct choice is (C).

10. The speed is the absolute value of the velocity. The speed is the greatest when the velocity is farthest from zero (above or below). This occurs at point D when the (negative) velocity has a larger absolute value than B where the velocity is greatest.

    The correct choice is (D).

11. The curve $y = x^2 + 1$ intersects the line $y = 5$ when $x^2 + 1 = 5$ or $x = \pm 2$.

    Since the area of the shaded region is in Quadrant I only, then

    $$A = \int_0^2 (5 - (x^2 + 1))\,dx = \int_0^2 (4 - x^2)\,dx = 4x - \left.\frac{x^3}{3}\right|_0^2$$

    $$= \left(8 - \frac{8}{3}\right) - (0 - 0) = 8 - \frac{8}{3} = \frac{16}{3}$$

    The correct choice is (B).

12. Recall that $\displaystyle\int \frac{du}{\sqrt{a^2 - u^2}} = \sin^{-1}\left(\frac{u}{a}\right) + C$

    Let $u = x$ and $a = 2$ with $du = dx$.

    Thus $\displaystyle\int \frac{dx}{\sqrt{4 - x^2}} = \int \frac{du}{\sqrt{a^2 - u^2}} = \sin^{-1}\left(\frac{u}{a}\right) + C$ or $\sin^{-1}\left(\dfrac{x}{2}\right) + C$

    Therefore, the correct choice is (A).

13. In general, a function $f(x)$ has a point of inflection where $f''(x) = 0$ or when $f''(x)$ is undefined, and the concavity changes at that point.

    In the given problem, $f'(x) = 4x - \frac{k}{x^2}$ and $f''(x) = 4 + \frac{2k}{x^3}$.

    Since $f(x)$ has a point of inflection at $x = -1$, then $4 + \frac{2k}{(-1)^3} = 4 - 2k = 0$

    or $k = 2$.

    The correct choice is (E).

14. Let $u = 3x + 4$

    $du = 3\,dx$ or $dx = \frac{1}{3}du$

    Thus $\int \sin(3x+4)\,dx = \frac{1}{3}\int \sin u\,du = \frac{1}{3}(-\cos(3x+4)) + C$ or $-\frac{1}{3}\cos(3x+4) + C$.

    The correct choice is (A).

15. A function $y = f(x)$ is concave downward when $f''(x) < 0$.

    In the given function $y = \frac{2}{4-x}$,

    $$y'' = \frac{4}{(4-x)^3}$$

    $x = 4$ is the only critical point

	$x < 4$	$x > 4$
y''	+	−

    Therefore, $y = \frac{2}{4-x}$ is concave downward when $x > 4$.

    The correct choice is (E).

16. $x(t) = \frac{1-t}{1+t}$

    $$v(t) = x'(t) = \frac{-(1+t)-(1-t)}{(1+t)^2} = \frac{-2}{(1+t)^2}$$

    $$a(t) = v'(t) = \frac{4+4t}{(1+t)^4} = \frac{4(1+t)}{(1+t)^4} = \frac{4}{(1+t)^3}$$

    The acceleration at $t = 0$ is $a(0) = \frac{4}{(1+0)^3} = 4$.

    The correct choice is (E).

17. Since $f(x) > g(x)$, the area is given by $\int_a^b f(x) - g(x)\,dx = \int_a^b [g(x) + 5 - g(x)]\,dx$

$$= \int_a^b 5\,dx$$

$$= 5x\Big|_a^b$$

$$= 5b - 5a$$

The correct choice is (C).

18. Carefully translate the sentence into symbols.

The rate of change of the surface area of a cube, S, with respect to time, t, is $\frac{dS}{dt}$, which is directly proportional (k times) to the square root of one-sixth of the surface area; $k\sqrt{\frac{S}{6}}$.

The correct choice is (C).

19. Refering to the accompanying diagram,

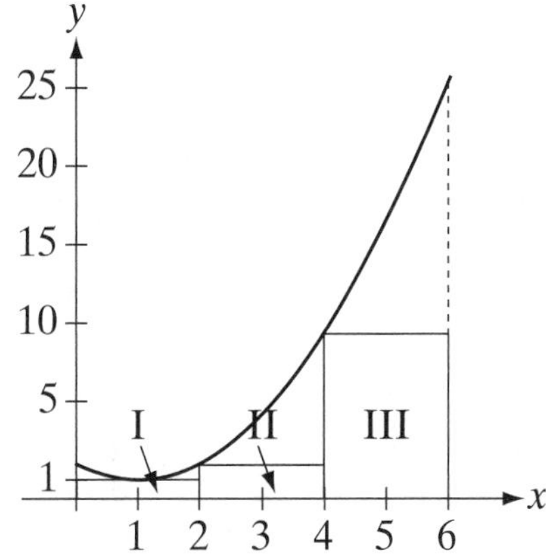

Area of Rectangle I: $2 \cdot f(1) = 2 \cdot 1 = 2$

Area of Rectangle II: $2 \cdot f(2) = 2 \cdot 2 = 4$

Area of Rectangle III: $2 \cdot f(4) = 2 \cdot 10 = 20$

Total Area of Rectangles: $2 + 4 + 20 = 26$

The correct choice is (B).

20. Notice that the segments that make up the slope field are always parallel across the page. This means that the x-coordinates do not affect the slope and therefore x does not appear in the differential equation. This eliminates choices (A), (B) and (C). The slope field for choice (D) can contain only segments with positive slopes. This eliminates (D).

    The correct choice is (E).

21. If, for all values of x, $f'(x) < 0$ and $f''(x) > 0$, then f is always decreasing and concave upward.

    The correct choice is (D).

22. The second derivative is found by using the Product Rule and the Chain Rule to differentiate $\frac{dy}{dx} = x^2y^2$:

$$\frac{d^2y}{dx^2} = x^2\left(2y\frac{dy}{dx}\right) + 2xy^2$$

    Then substitute for $\frac{dy}{dx}$:

$$\frac{d^2y}{dx^2} = x^2(2y(x^2y^2)) + 2xy^2 = 2x^4y^3 + 2xy^2$$

    The correct choice is (E).

23. By the quotient rule the integrand $\frac{f(x)g'(x) - g(x)f'(x)}{(f(x))^2} = \frac{d}{dx}\left(\frac{g(x)}{f(x)}\right)$. Therefore

$$\begin{aligned}\int_2^8 \frac{f(x)g'(x) - g(x)f'(x)}{(f(x))^2}\,dx &= \int_2^8 \frac{d}{dx}\left(\frac{g(x)}{f(x)}\right)dx \\ &= \left.\frac{g(x)}{f(x)}\right|_2^8 \\ &= \frac{g(8)}{f(8)} - \frac{g(2)}{f(2)} \\ &= \frac{-6}{3} - \frac{2}{1} = -4\end{aligned}$$

    The correct choice is (E).

24. By the Product Rule $h'(x) = f(x)\,g'(x) + g(x)\,f'(x)$ and

$h'(3) = f(3)\,g'(3) + g(3)\,f'(3) = (1)(1) + (3)\left(-\frac{1}{3}\right) = 0$

The values of the functions are read from the graph; the values of the derivatives are the slopes of the lines at $x = 3$.

The correct choice is (C).

25. $C = 2\pi r$ and therefore $\frac{dC}{dt} = 2\pi\frac{dr}{dt}$. Since $\frac{dC}{dt} = 0.5, \frac{dr}{dt} = \frac{0.5}{2\pi} = \frac{1}{4\pi}$ meters/minute.

$A = \pi r^2$ and $\frac{dA}{dt} = 2\pi r\frac{dr}{dt}$. When $r = 4$ meters, $\frac{dA}{dt} = 2\pi(4)\frac{1}{4\pi} = 2\ m^2$/min.

The correct choice is (A).

26. $\int_{-2}^{1} x\,f(x)\,dx = \int_{-2}^{0} x(x)\,dx + \int_{0}^{1} x(x+1)\,dx = \frac{x^3}{3}\bigg|_{-2}^{0} + \left(\frac{x^3}{3} + \frac{x^2}{2}\right)\bigg|_{0}^{1} = \frac{7}{2}$

The correct choice is (D).

27. The average value is $\frac{1}{0-(-4)}\int_{-4}^{0} \cos\left(\frac{1}{2}x\right)\,dx = \frac{1}{4}\int_{-4}^{0} \cos\left(\frac{1}{2}x\right)\,dx$.

Let $u = \frac{1}{2}x,\ du = \frac{1}{2}dx \Rightarrow dx = 2du$.

Since $\int \cos\left(\frac{1}{2}x\right)\,dx = 2\int \cos(u)\,du = 2\sin\left(\frac{1}{2}x\right) + C$

Then, $\frac{1}{4}\int_{-4}^{0} \cos\left(\frac{1}{2}x\right)\,dx = \frac{1}{4}\cdot 2\sin\left(\frac{1}{2}x\right)\Big|_{-4}^{0} = \frac{1}{2}\sin\left(\frac{1}{2}x\right)\Big|_{-4}^{0} = \frac{1}{2}(0 - \sin(-2)) = -\frac{1}{2}\sin(-2)$

Since $\sin(-2) = -\sin(2)$, then $-\frac{1}{2}\sin(-2) = -\frac{1}{2}(-\sin(2)) = \frac{1}{2}\sin(2)$.

The correct choice is (E).

28. If $u = \sqrt{2x}$, then $du = \frac{2}{2\sqrt{2x}}dx$. Thus, $du = \frac{dx}{u}$ and $dx = u\,du$.

Adjusting the limits of integration, when $x = 2, u = 2$ and when $x = 8, u = 4$. Making all these substitutions $\int_{2}^{8} \frac{dx}{\sqrt{2x} + 1} = \int_{2}^{4} \frac{u\,du}{u+1}$.

The correct choice is (D).

29. The graph of the region described is:

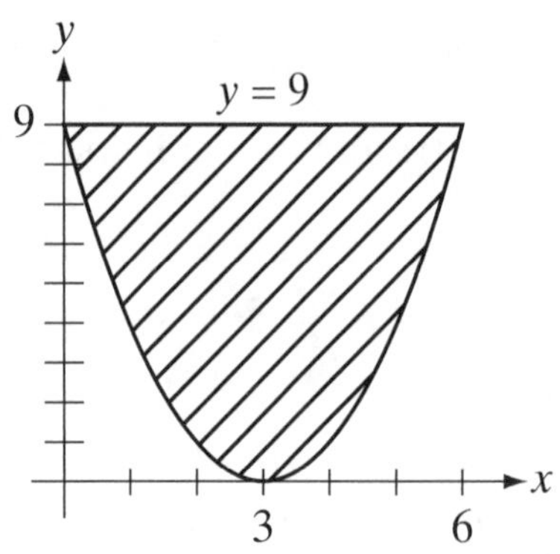

Using the disk method, the radius is $9 - (x - 3)^2$ and the volume equals

$$\pi \int_0^6 (9 - (x - 3)^2)^2 dx$$

Since the parabola $y = (x - 3)^2$ is symmetric about the line $x = 3$, the volume could then be given as: $2\pi \int_0^3 (9 - (x - 3)^2)^2 dx$

The correct choice is (B).

30. The tangent lines will be parallel when the derivatives of f and g are equal.

$$\begin{aligned} f'(x) &= g'(x) \\ \sec^2 x &= 2x \\ \frac{1}{\cos^2(x)} &= 2x \\ x &\approx 2.083 \end{aligned}$$

Use the solver feature on a calculator or find the intersection of the two functions..

The correct choice is (C).

31. The average rate of change of f is given by $\frac{\Delta f}{\Delta t} = \frac{\frac{1}{b} - \frac{1}{a}}{b - a} = \frac{a - b}{ab(b - a)} = -\frac{1}{ab}$. The derivative is $f'(t) = -\frac{1}{t^2}$. Set these equal to each other and solve for t:

$$\begin{aligned} -\frac{1}{ab} &= -\frac{1}{t^2} \\ t^2 &= ab \\ t &= \sqrt{ab} \end{aligned}$$

The correct choice is (B).

32. The number of subintervals in this situation does not matter. To compare the relative sizes draw one rectangle or trapezoid on the interval $[a,b]$.

Since the function is increasing, the left sum will be the least of the four numbers and the right sum will be the greatest. Since $T = \frac{1}{2}(L + R)$, the trapezoidal sum is always between the left and right sums.

Since the function is concave upwards, the top of the trapezoids will lie above the graph and therefore the area of the trapezoids will be greater than the integral which represents the area between the graph and the x-axis. Therefore, the correct order is $L < A < T < R$.

The correct choice is (A).

33. $R(t)$ is the rate of change of the volume of the water and therefore a derivative. Thus by the Fundamental Theorem of Calculus, $\int_0^3 R(t)\,dt$ represents the total amount of water that leaks out in the first three hours.

Keep in mind that the definite integral of a rate of change gives total change; the integral of the rate at which water leaks is the amount of water that leaked.

The correct choice is (C).

34. $h(x) = f(g(x))$

By the Chain Rule, $h'(x) = f'(g(x)) \cdot g'(x)$

$$
\begin{aligned}
\text{Therefore, } h'(1) &= f'(g(1)) \cdot g'(1) \\
&= f'(3) \cdot g'(1) \quad \text{since } g(1) = 3 \\
&= (-5)(-3) \quad \text{since } f'(3) = -5 \text{ and } g'(1) = -3 \\
&= 15
\end{aligned}
$$

The correct choice is (B).

35. Since f is increasing, $\dfrac{f(b) - f(0)}{b - 0} > 0$. Since this is the value of $f'(c)$, $f'(c) > 0$. (D) is true and (C) is false.

(A) is false: $f''(c) < 0$ may be true, but only if the function is concave down. (B) is false since it assumes $f(0) = 0$. (E) is false since this implies that the function is not always increasing.

The correct choice is (D).

36. Graph $y_1 = e^{(x^2)} - 2$ and $y_2 = \sqrt{4 - x^2}$ in the viewing rectangle $[-3,3]$ by $[-3,3]$. The points of intersection of y_1 and y_2 are $x = \pm 1.1373$. (These two functions are even functions, i.e. are symmetric with respect to the y-axis. Consequently, finding one point of intersection, e.g. $x = 1.1373$, will yield the other point of intersection, $x = -1.1373$).

Since $y_2 > y_1$ on $[-1.1373, 1.1373]$, the area is represented by

$$A = \int_{-1.1373}^{1.1373} [\sqrt{4 - x^2} - (e^{(x^2)} - 2)]\, dx = \int_{-1.1373}^{1.1373} (\sqrt{4 - x^2} - e^{(x^2)} + 2)\, dx = 5.050$$

The correct choice is (D).

37. Because of the symmetry in the acceleration values, the left-hand, right-hand and trapezoidal approximations give the same value. The left Riemann sum approximation is

$$2(2) + 3(2) + 4(2) + 3(2) = 24$$

This is the accumulated net change in velocity. Add this amount to the initial velocity to find the final velocity: $24 + 4 = 28$ feet/sec.

The correct choice is (E).

38. $\int_0^7 f(x)\, dx = \int_0^1 f(x)\, dx + \int_1^7 f(x)\, dx$

$1 = 5 + \int_1^7 f(x)\, dx \Rightarrow \int_1^7 f(x)\, dx = -4$

Since even functions are symmetric to the y-axis, the graph between –7 and –1 encloses the same area as between 1 and 7 and has the same amounts above and below the x-axis, therefore

$\int_{-7}^{-1} f(x)\, dx = \int_1^7 f(x)\, dx = -4$

The correct choice is (B).

39. The local maximums and minimums of $f'(x)$ correspond to points of inflection on the graph of function $f(x)$. For example, at the top left maximum point, $f'(x)$ changes from increasing to decreasing, thus the function $f(x)$ changes from concave up to concave down. Another way to find the point of inflection is that at the extreme points of $f'(x)$, the derivative's derivative of $f(x)$, the second derivative of $f(x)$ changes sign, indicating a point of inflection. There are four such points shown in the graph of $f'(x)$.

The correct choice is (E).

40. The average value of a function on an interval $[a,b]$ is given by $\frac{1}{b-a}\int_a^b f(x)\,dx$. The integral gives the area under the graph. So in this problem the average value is $\frac{area}{12} = \frac{4 \cdot 8 + \frac{1}{4}\pi(4^2)}{12} \approx 3.714$

The correct choice is (B).

41. Using a calculator, evaluate the definite integral:

By the Fundamental Theorem of Calculus or

$$\begin{aligned} \int_1^5 \sqrt[3]{x^2+4x}\,dx &\approx 10.882 \\ g(5) - g(1) &\approx 10.882 \\ \text{or } 7 - 10.8822 &\approx g(1) \\ -3.882 &\approx g(1) \end{aligned}$$

The correct choice is (A).

42. The rate of production is given by the first derivative of A: $A'(t) = 48 - 12(t-3)^2$.

The maximum increase in the rate of production occurs when this expression has its maximum value, that is, when $A''(t) = 0$.

$A''(t) = -24(t-3) \Rightarrow A''(t) = 0$ when $t = 3$. Three hours after 8:00 am is 11:00 am.

Note: $A'''(3) = -24$ indicating that $t = 3$ is a maximum (Second Derivative Test). Or use the First Derivative Test: A'' switches from positive to negative at $t = 3$, so $t = 3$ is a maximum.

The correct choice is (C).

43. $f'(0) > 0$ and $f'(6) > 0$ since the function is increasing for $x \le 8$.

$f'' < 0$ since the graph is concave down for $x \le 10$.

$f''(10) = 0$ since this is a point of inflection.

$f''(12) > 0$ since the graph is concave up for $x \ge 10$.

Therefore, the smallest value is $f''(4)$.

The correct choice is (C).

44. The limiting value is 6000 as time goes on, i.e. $\lim_{t\to+\infty}(6000-5500e^{-0.159t})=6000$. Using a calculator, find the intersection of $P(t)$ and $P=3000$. Another possibility is to solve the equation

$$3000 = 6000 - 5500e^{-0.159t}$$
$$-3000 = -5500e^{-0.159t}$$
$$\frac{30}{55} = e^{-0.159t}$$
$$\ln\left(\frac{30}{55}\right) = -0.159t \Rightarrow t = \frac{\ln\left(\frac{30}{55}\right)}{-0.159} \Rightarrow t \approx 3.8$$

Thus the population will reach half of its limiting value in the <u>fourth</u> year.

The correct choice is (C).

45. Since $f'(x)>0$ for $0<x<6$; and $f'(x)<0$ for $x>6$, $f(x)$ increases for $0<x<6$ and $f(x)$ decreases for $x>6$. Obviously, the maximum point on the graph of $f(x)$ is at $x=6$.

Therefore, the maximum value of $f(x)$ is $f(6)$. To find $f(6)$, use the Fundamental Theorem of Calculus:

$$\int_2^6 f'(x)\,dx = f(6)-f(2) \text{ or, } f(6)=f(2)+\int_2^6 f'(x)\,dx$$

Since it is given that $f(2)=10$, then $f(6)=10+\int_2^6 f'(x)\,dx$.

Now, $\int_2^6 f'(x)\,dx$ is the area under $f'(x)$ from $x=2$ to $x=6$, which is approximately 100. Note that each square box has an area of 5, and there are approximately 20 square boxes combined.

Therefore, the maximum value of $f(x)\approx 10+100\approx 110$.

The correct choice is (E).

1a. $L'(x) = (167.5)\left(\frac{2\pi}{366}\right)\cos\left(\left(\frac{2\pi}{366}\right)(d-80)\right)$.

The maximum occurs when the derivative is zero, which is when $\cos\left(\left(\frac{2\pi}{366}\right)(d-80)\right) = 0$.

Hence $\left(\frac{2\pi}{366}\right)(d-80) = \frac{\pi}{2}$ or $\frac{3\pi}{2}$, which means $d = \frac{\pi}{2}\frac{366}{2\pi} + 80 = 171.5$

Or $d = \frac{3\pi}{2}\frac{366}{2\pi} + 80 = 354.5$

The maximum occurs on the 172nd day. (Day 355 is in December and is the minimum.)

1b. The total number of minutes of sunlight is given by $\int_0^{366} L(d)\,dt = 267{,}546$ minutes. This may also be found by multiplying the average minutes of sunlight (731) by the number of days (366).

1c. The mean value is given by $\frac{1}{366}\int_0^{366} L(d)\,dt = 731$, by use of a calculator.

OR: $L(d)$ is a sine function which has the same area above and below the horizontal line $y = 731$ therefore its average value is 731. The graph of $L(d)$ and $y = 731$ are shown below in the window $x[0,366]$ by $y[500,1000]$

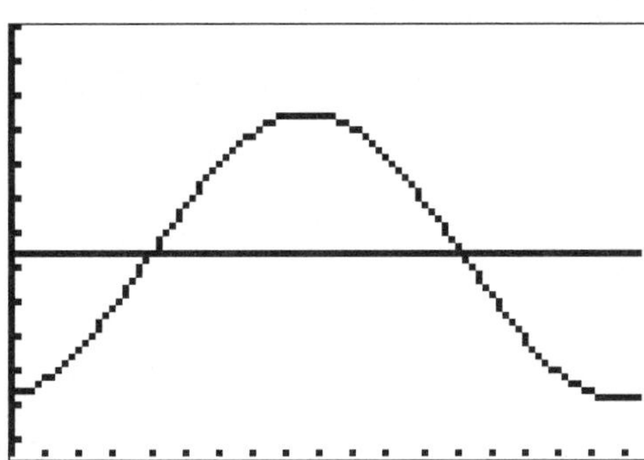

2a. At any point the slope of the hill is given by $\frac{d}{dx}f(x) = -\frac{50}{100}\sin\left(\frac{x}{100}\right)$. An equation of the tangent line at $x = a$ is:

$$y - 50\cos\left(\frac{a}{100}\right) = -\frac{50}{100}\left(\sin\left(\frac{a}{100}\right)\right)(x-a)$$

$$y = -\frac{1}{2}\left(\sin\left(\frac{a}{100}\right)\right)x + \frac{a}{2}\sin\left(\frac{a}{100}\right) + 50\cos\left(\frac{a}{100}\right)$$

2b. The top of the hill is at (0,50) and the person's eyes are at (0,55), which is the y-intercept of the line. Therefore,

$$\frac{a}{2}\sin\left(\frac{a}{100}\right) + 50\cos\left(\frac{a}{100}\right) = 55$$

Solving this equation on your calculator, the first point of the person's line of vision and the hill occurs at $a = 45.935$. Extending this line of vision, the person's far point of sight of the valley would be at $a = 398.341$, which obviously is past the middle of the valley. (Note that the other solution to the above equation, $a = 226.571$, is extraneous to the problem since it would imply that the person was able to see through the dense hill).

2c. The middle of the valley is at $x = 100\pi$ (half the period of $f(x) = 50\cos\left(\frac{x}{100}\right)$).

$f(100\pi) = -50$ while $y(100\pi) = -\frac{1}{2}\left(\sin\left(\frac{45.935}{100}\right)\right)(100\pi) + 55 = -14.643$.

Since $25 < -14.643 - (-50)$ the top of the flagpole is at $(100\pi - 25)$, which is below the person's line of sight along the tangent line. Hence the top of the flagpole cannot be seen by the person.

REMINDER: Numeric and algebraic answers need not be simplified, and if simplified incorrectly credit will be lost.

3a. $x(3) = 7(3) - 4(3^2) + \int_0^3 s^2\, ds = 21 - 36 + \left.\frac{s^3}{3}\right|_0^3 = 21 - 36 + 9 = -6$

$v(t) = x'(t) = 7 - 8t + t^2$

$v(3) = 7 - 8(3) + 3^2 = -8$

3b. Speed $= |v(3)| = 8.$

$a(t) = v'(t) = -8 + 2t$

$a(3) = -2$

Since the velocity and acceleration have different signs, the speed is decreasing.

3c. $v(t) = 7 - 8t + t^2 = (t-7)(t-1) = 0$

For $0 < t < 1$, $v(t) < 0$, and for $1 < t < 4$, $v(t) > 0$. Hence there is a change in direction of the particle at $t = 1$.

(Note: Although $v(7) = 0$, $t = 7$ is outside of the given domain and therefore not considered as a solution value.)

3d. Since the particle changes direction only once, check the function's value at that point and also at the endpoints of the given domain.

$x(0) = 0$

$x(1) = 7(1) - 4(1^2) + \int_0^1 s^2\, ds = 7 - 4 + \left.\frac{s^3}{3}\right|_0^1 = 7 - 4 + \frac{1}{3} = \frac{10}{3}$

$x(4) = 7(4) - 4(4^2) + \int_0^4 s^2\, ds = 7 - 4 + \left.\frac{s^3}{3}\right|_0^4 = 28 - 64 + \frac{64}{3} = -\frac{44}{3}$

The particle is farthest right at $t = 1$ and farthest left at $t = 4$.

4a. By the Chain rule

$$\begin{aligned} h(1) &= f(g(1)) - g(f(1)) \\ &= f(2) - g(3) \\ &= 5 - 5 \\ &= 0 \\ h(4) &= f(g(4)) - g(f(4)) \\ &= f(5) - g(2) \\ &= 4 - 4 \\ &= 0 \end{aligned}$$

Therefore by Rolle's Theorem (or the Mean Value Theorem) there must exist a number c with $1 < c < 4$, such that $h'(c) = 0$.

4b.
$$\begin{aligned} h'(x) &= f'(g(x))\,g'(x) - g'(f(x))\,f'(x) \\ h'(3) &= f'(g(3))\,g'(3) - g'(f(3))\,f'(3) \\ 0 &= 8g'(3) - (9)(5) \\ \frac{45}{8} &= g'(3) \end{aligned}$$

4c.
$$\begin{aligned} w(x) &= 7 + \int_1^{f(x)} g(t)\,dt \\ w(3) &= 7 + \int_1^{f(3)} g(t)\,dt = 7 + \int_1^{1} g(t)\,dt = 7 + 0 = 7 \\ w'(x) &= 0 + g(f(x))\,f'(x) \\ w'(3) &= g(f(3))\,f'(3) \\ &= 2 \cdot 5 = 10 \end{aligned}$$

The line through (3,7) with a slope of 10 is $y - 7 = 10(x - 3)$ or $y = 10x - 23$.

5a.

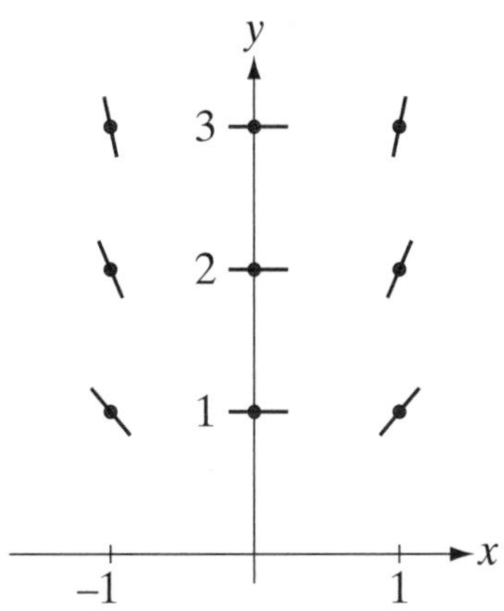

5b. Solve the differential equation by separating the variables and integrating both sides.

$$y^{-2}dy = x\,dx$$

$$\frac{-1}{y} = \frac{x^2}{2} + C$$

$$y = \frac{-1}{\frac{x^2}{2} + C}$$

5c. Substitute the initial condition:

$$\frac{-1}{0^2 + C} = 1$$

$$C = -1$$

$$y = \frac{-1}{\frac{x^2}{2} - 1} = \frac{-2}{x^2 - 2}$$

5d. The functions will have vertical asymptotes when the denominator has real roots. This will occur when $C \leq 0$.

6a. The function, f, is concave down when its derivative is decreasing. This occurs on the interval $[-1,1.5]$. On this interval, $G'' \leq 0$.

Note: $G'(x) = f(x)$ and $G''(x) = f'(x)$

Since $G'(x) = \frac{d}{dx}\int_{-2}^{x} f(t)\,dt = f(x)$

6b. $m = 2$, $G(0) = 7$. From the graph, $G'(0) = f(0) = 2$ is the slope of the tangent line. The point $(0,7)$ is on the tangent line at $x = 0$, so this point must also be on the graph of G. Therefore $G(0) = 7$.

6c. Since the average value of f on $0 \leq x \leq 3$ is $\frac{1}{3-0}\int_{0}^{3} f(x)\,dx = 0$, $\int_{0}^{3} f(x)\,dx = 0$, and

$$\begin{aligned} G(0) &= 7 \\ G(3) &= G(0) + \int_{0}^{3} f(x)\,dx \\ &= 7 + 0 \\ &= 7 \end{aligned}$$

Note: $G(3) = G(-2) + \int_{-2}^{3} f(t)\,dt$

$G(0) = G(-2) + \int_{-2}^{0} f(t)\,dt$

Therefore $G(3) - G(0) = \int_{-2}^{3} f(t) + \int_{-2}^{0} f(t)\,dt = \int_{0}^{3} f(t)\,dt$

$G(3) = G(0) + \int_{0}^{3} f(t)\,dt = 7 + 0 = 7$

Sample Examination II

1. $$\int_0^2 (2x^3+3)\,dx = \left(\frac{2x^4}{4}+3x\right)\Bigg|_0^2$$
$$= \left(\frac{x^4}{2}+3x\right)\Bigg|_0^2$$
$$= (8+6)-(0+0) = 14$$

The correct choice is (C).

2. The average value of f' is given by $\frac{1}{2-(-2)}\int_{-2}^{2} f'(x)\,dx$. Using the Fundamental Theorem of Calculus, $\frac{1}{2-(-2)}\int_{-2}^{2} f'(x)\,dx = \frac{f(2)-f(-2)}{4} = \frac{e-a}{4}$

The correct choice is (B).

3. This question is testing knowledge of the Intermediate Value Theorem. In this problem, it is given that $f(-5)=3$ and $f(-1)=-2$. Since $f(x)=0$ for only one value of x, then x has to be between –5 and –1. The only number among the choices that satisfies this is –2.

The correct choice is (B).

4. $$f(x) = (2+3x)^4$$
$$f'(x) = 4(2+3x)^3\,(3)$$
$$f''(x) = 3\cdot 4(2+3x)^2\,(3)^2$$
$$f'''(x) = 2\cdot 3\cdot 4(2+3x)(3)^3$$
$$f^{(4)}(x) = 1\cdot 2\cdot 3\cdot 4\cdot (3)^4 = 4!\,(3^4)$$

If $y=(a+bx)^n$, then the nth derivative, $\frac{d^n y}{dx^n} = n!\,(b^n)$

The correct choice is (C).

5. $f(x) = x^4 - 8x^2$

$f'(x) = 4x^3 - 16x = 4x(x^2 - 4)$

$f'(x) = 0$ when $x = 0, -2,$ and 2.

	$x < -2$	$-2 < x < 0$	$0 < x < 2$	$x > 2$
f'	$-$	$+$	$-$	$+$

$f(x)$ has a relative minimum where $f'(x)$ switches sign from – to +.

Therefore, $f(x)$ has a relative minimum at $x = 2$ and $x = -2$ only.

The correct choice is (D).

6. $$\int \sqrt{x}\,(x+2)\,dx = \int \left(x^{\frac{3}{2}} + 2x^{\frac{1}{2}}\right) dx$$
$$= \frac{2}{5}x^{\frac{5}{2}} + 2 \cdot \frac{2}{3}x^{\frac{3}{2}} + C$$
$$= \frac{2}{5}x^{\frac{5}{2}} + \frac{4}{3}x^{\frac{3}{2}} + C$$

The correct choice is (D).

7. The $\lim\limits_{h \to 0} \dfrac{|x+h| - |x|}{h}$ at $x = 3$, is the definition of the derivative of the absolute value function at $x = 3$.

If $f(x) = |x|$, and $x > 0$, then $f(x) = x$ and $f'(x) = 1$.

Therefore, $f'(3) = 1$.

(<u>Note</u>: Since $|x| = -x$ when $x < 0$, $f'(-3) = -1$)

The correct choice is (C).

8. If the function $y = x^4 + bx^2 + 8x + 1$ has a horizontal tangent for some value of x, then the slope of y, y', is equal to 0 for that value of x.

$$y' = 4x^3 + 2bx + 8 \Rightarrow 4x^3 + 2bx + 8 = 0$$

If the function y has a point of inflection for some value of x, then $y'' = 0$ for that value of x.

$$y'' = 12x^2 + 2b \Rightarrow 12x^2 + 2b = 0 \Rightarrow 2b = -12x^2$$

Since the function has a horizontal tangent and a point of inflection for the same value of x,

$$4x^3 + 2bx + 8 = 12x^2 + 2b$$

Substituting $-12x^2$ for $2b$,

$$\begin{aligned} 4x^3 + (-12x^2)(x) + 8 &= 12x^2 + (-12x^2) \\ 4x^3 - 12x^3 + 8 &= 0 \\ -8x^3 &= -8 \\ x^3 &= 1 \\ x &= 1 \end{aligned}$$

Since $x = 1$ and $2b = -12x^2$, then $2b = -12$, $b = -6$.

Therefore the value of b is -6.

The correct choice is (A).

9. Horizontal asymptotes are the graphical manifestations of limits at ∞ and $-\infty$.

$\lim\limits_{x\to-\infty} \dfrac{4e^x + 7}{e^x - 1} = -7$ and $\lim\limits_{x\to\infty} \dfrac{4e^x + 7}{e^x - 1} = \lim\limits_{x\to\infty} \dfrac{4 + 7e^{-x}}{1 - e^{-x}} = 4$. Therefore the asymptotes are $y = -7$ and $y = 4$.

The correct choice is (D).

10. Method 1: The average rate of change is given by $\dfrac{f(b) - f(a)}{b - a} = \dfrac{e^{(9)} - e^{(9)}}{3 - (-3)} = 0$. The instantaneous rate of change is the derivative $f'(x) = 2xe^{(x^2)}$. Solving $2xe^{(x^2)} = 0$ gives only one value of $x = 0$.

Method 2: The instantaneous rate of change is given by the derivative of f: $f'(x) = 2xe^{(x^2)}$. Since the derivative is always increasing $\left(f''(x) = (4x^2 + 2)\, e^{(x^2)} > 0\right)$ it will equal every real number exactly once. Therefore whatever the average value is, it will equal it once.

The correct choice is (B).

11. According to the Mean Value Theorem, $f'(c) = \dfrac{f(b) - f(a)}{b - a}$.

$f(x) = x^3 \Rightarrow f'(c) = 3c^2$

So $3c^2 = \dfrac{8 - (-1)}{2 - (-1)} \Rightarrow 3c^2 = 3 \Rightarrow c = \pm 1$

The value of $c = -1$ is to be rejected, since the Mean Value Theorem guarantees a value of c, $a < c < b$; therefore $-1 < c < 2 \Rightarrow c = 1$ only.

The correct choice is (B).

12. Calculate the values and arrange them in order. The values are calculated by adding and subtracting the areas between the graph and the x-axis. Since the lower limit of integration is 1, the areas of the regions to the left of $x = 1$ are given the opposite sign.

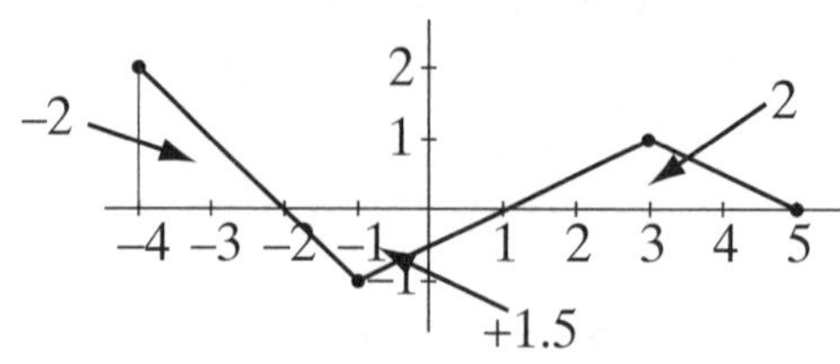

$g(-4) = (1.5 - 2) = -0.5$

$g(-2) = 1.5$

$g(1) = 0$

$g(5) = 2$

Arrange these in order $g(-4) < g(1) < g(-2) < g(5)$

The correct choice is (A).

13. $x + y = xy$

By implicit differentiation, $1 + \frac{dy}{dx} = x\left(\frac{dy}{dx}\right) + y$

Solving for $\frac{dy}{dx}$,

$$1 - y = x\left(\frac{dy}{dx}\right) - \frac{dy}{dx}$$

$$1 - y = \frac{dy}{dx}(x - 1)$$

$$\frac{dy}{dx} = \frac{1 - y}{x - 1}$$

The correct choice is (C).

14. Let $u = g(x)$ and $du = g'(x)\,dx$, then $\int_2^4 f'(g(x))\,g'(x)\,dx = \int f'(u)\,du = f(u) + C$

Thus $f(g(x))\Big|_2^4 = f(g(4)) - f(g(2))$

The correct choice is (D).

15. Integrate the velocity function $v(t)$ to find the position function $y(t)$:

$$y(t) = \int v(t)\,dt = \int (8 - 2t)\,dt = 8t - t^2 + C$$

The velocity is positive when $t < 4$ since during this time the particle is moving upwards.

At $t = 4$, $v(4) = 0$ and the particle stops moving up when it reaches the origin and begins moving down, therefore $y(4) = 0$.

$$y(4) = 8(4) - 4^2 + C = 0 \Rightarrow C = -16$$

Therefore, the position function is $y(t) = -t^2 + 8t - 16$.

The correct choice is (A).

16. Using the substitution $u = \sqrt{x-1}$

$$u^2 = x - 1 \Rightarrow x = u^2 + 1 \text{ and } dx = 2u\,du$$

Use $u = \sqrt{x-1}$ to change the limits of integration,

thus when $x = 2, u = 1$, and when $x = 5, u = 2$.

So $\displaystyle\int_2^5 \frac{\sqrt{x-1}}{x}\,dx = \int_1^2 \frac{u}{u^2+1}\cdot 2u\,du = \int_1^2 \frac{2u^2}{u^2+1}\,du$

The correct choice is (E).

17. A function $f(x)$ is increasing when $f'(x) > 0$.

$f'(x) = 3x^2 + 12x + 9 = 3(x+1)(x+3)$

$f'(x) = 0$ when $x = -3, -1$.

	$x < -3$	$-3 < x < -1$	$x > -1$
f'	+	−	+

Therefore, $f(x)$ is increasing when $x \le -3$ or $x \ge -1$.

The correct choice is (D).

18. $\int_0^2 (2x^3 - kx^2 + 2k)\,dx = \frac{2x^4}{4} - \frac{kx^3}{3} + 2kx\Big|_0^2 = \left(\frac{32}{4} - \frac{8k}{3} + 4k\right) - (0) = 8 - \frac{8k}{3} + 4k$

Since $\int_0^2 (2x^3 - kx^2 + 2k)\,dx = 12$

Therefore, $8 - \frac{8k}{3} + 4k = 12$

$$24 + 4k = 36$$
$$k = 3$$

The correct choice is (E).

19. Let $u = \tan x$ and $du = \sec^2 x\,dx$.

$\int (\sec^2 x)(\tan^2 x)\,dx = \int u^2 du = \frac{u^3}{3} + C$ or $\frac{\tan^3 x}{3} + C$

The correct choice is (A).

20. In general, the derivative of $y = \ln u$ is $y' = \frac{du}{u}$.

The derivative of $y = \ln \sqrt{1 - x^2}$ is

$y' = \frac{1}{\sqrt{1-x^2}} \cdot d(\sqrt{1-x^2}) = \frac{1}{\sqrt{1-x^2}} \cdot \frac{1}{2} \cdot \frac{1}{\sqrt{1-x^2}} \cdot -2x = \frac{-x}{1-x^2}$

By multiplying both numerator and denominator by –1, the expression $\frac{-x}{1-x^2} = \frac{x}{x^2-1}$. The derivative of $y = \ln \sqrt{x^2 - 1}$ is $y' = \frac{x}{x^2-1}$.

The correct choice is (B).

21. A function whose derivative is a constant multiple of itself is equivalent to the expression $y' = ky$ (where k is the constant). The general solution of $y' = ky$ is $y = Ce^{kt}$. Therefore, the function must be exponential.

The correct choice is (C).

22. The graph is concave downward when $y'' < 0$.

$y = x^3 - 6x^2$, $y' = 3x^2 - 12x$, $y'' = 6x - 12$, $y'' < 0$ when $6x - 12 < 0$ or $x < 2$

Hence, $y = x^3 - 6x^2$ is concave downward when $x < 2$.

The correct choice is (C).

23. The volume of the cube is $V = s^3$. Differentiating with respect to time gives $\frac{dV}{dt} = 3s^2 \frac{ds}{dt}$.

Since $\frac{ds}{dt} = 0.05$, $\frac{dV}{dt} = 3s^2(0.05) = 0.15s^2$.

The correct choice is (D).

24.

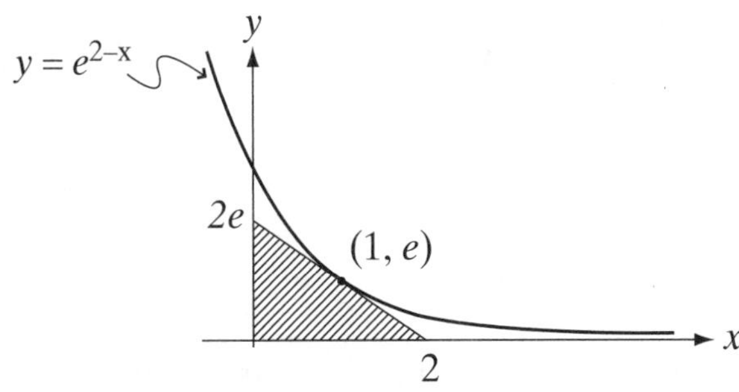

First, find the equation of the tangent line to $y = e^{2-x}$ at $(1, e)$. $y' = -e^{2-x}$; therefore at $x = 1$, $y'(1) = -e$. The equation of the tangent line is $y - e = -e(x - 1)$ or $y = -ex + 2e$. To find the area of the triangle (the shaded region) calculate the base and height of the triangle, which are the x-intercept and y-intercept of the tangent line. The x-intercept is $x = 2$ and the y-intercept is $y = 2e$.

Therefore, the area of the triangle $= \frac{1}{2}(\text{base})(\text{height}) = \frac{1}{2}(2)(2e) = 2e$.

The correct choice is (A).

25. The graph of the function g will have points of inflection wherever its *second* derivative changes sign. Since $g' = f$ and $g'' = f'$, the second derivative of g will change sign where the *first* derivative of f changes from positive to negative or from negative to positive. This occurs three times on the given interval. Hence there are three points of inflection on the graph of g. These three points of inflection also correspond to the three max/min points of f.

The correct choice is (D).

26. $y = \sqrt[3]{x^2 - 1} = (x^2 - 1)^{\frac{1}{3}}$

$y' = \frac{1}{3}(x^2 - 1)^{-\frac{2}{3}}(2x)$

$y'(3) = \left(\frac{1}{3}\right)\left(\frac{1}{4}\right)(6) = \frac{1}{2}$

Thus the normal line to the curve will have a slope of –2.

The equation of the normal line to the curve $y = \sqrt[3]{x^2 - 1}$ at (3,2)

is $y - 2 = -2(x - 3)$ or $y + 2x = 8$

The correct choice is (D).

27. Since the velocity, $s'(t)$, and the acceleration, $s''(t)$, are both positive, the graph of $s(t)$ will be increasing $(s'(t) > 0)$ and concave upwards $(s''(t) > 0)$. Only the graph of (C) satisfies both of these conditions.

The correct choice is (C).

28. The maximum value of a function occurs where its first derivative changes from positive to negative or at an endpoint of the domain of the function. By the Fundamental Theorem of Calculus $g'(x) = f(x)$. From the graph $f(x)$ changes from positive to negative at $x = 3$. Thus the maximum value of $g(x)$ occurs at $x = 3$.

Alternate Method: Evaluate $g(x)$ as area of the regions of $f(t)$ to the x-axis. Those regions below the x-axis are counted as negative values.

Then $g(-3) = \int_0^{-3} f(t)\,dt = -8,$

$g(3) = \int_0^{3} f(t)\,dt = 3.5,$

and $g(5) = \int_0^{5} f(t)\,dt = 1.5$

Thus the maximum value of $g(x)$ occurs at $x = 3$.

The correct choice is (D).

29. Since $\lim_{x \to a} \frac{f(x) - f(a)}{x - a} = 0$, then $f'(a) = 0$ by definition of the derivative.

The correct choice is (C).

30. The graph of $f'(x)$ is always increasing, thus $f''(2) > 0$.

Since the derivative changes from negative to positive at $x = 2$, $f(2)$ is a minimum value. Since $f(1) = -2$ and $f(2) < f(1)$, then $f(2) < -2$.

From the graph, $f'(2) = 0$.

So arranging the numbers in order $f(2) < f'(2) < f''(2)$.

The correct choice is (A).

31. Since $f'(x)$ is always increasing and changes sign from negative to positive at $x = 0$, the First Derivative Test tells us that $x = 0$ is a relative minimum. (A) must be true.

Since $f'(x)$ is always increasing, $f(x)$ is always concave upwards and has no point of inflection. Hence, (B) and (C) are not true.

Since $f(0)$ is not given it cannot be determined if $f(x)$ passes through the origin — therefore, (D) need not be true.

Since the derivative $f'(x)$ is <u>not</u> even $((f'(10) \neq f'(-10))$, then $f(x)$ cannot be odd. The derivative of an odd function must be even. So (E) is not true.

The correct choice is (A).

32. Using a calculator, the amount is found by integrating $M(t)$: $\int_0^3 (4 - (\sin t)^3)\, dt \approx 10.667$.

The correct choice is (B).

33. To the right of the origin the function is decreasing; therefore, its derivative $f'(x) < 0$, and I is true. As $x \to +\infty$ and $x \to -\infty$, the graph flattens out so its slope $f'(x)$ is approaching zero; II is true and III is false.

The correct choice is (D).

34. Since $g'(x) = \frac{d}{dx}\int_a^x f(t)\, dt = f(x)$ and $f'(x) = g''(x)$, the correct graph can be identified by considering the concavity of g. From left to right, g is concave down, concave up, concave down. Thus the graph of $f'(x)$ changes from negative to positive to negative.

The correct choice is (B).

35. The acceleration becomes negative when the velocity stops increasing and starts decreasing (even though the velocity may still be positive). This corresponds to a point of inflection on the graph, where the concavity changes from up (on the left) to down (on the right). Point C appears to be the closest point where there is a point of inflection.

The correct choice is (C).

36. Graph $f'(x)$ in the window $[0,5]$ by $[-50,5]$. The derivative changes from negative (indicating a decreasing function) to positive (increasing function) at $x = 0.618$, thus indicating a relative minimum. Now check how the endpoint minimum at $x = 5$ compares to this value. This may be done by evaluating $\int_{0.618}^{5} [e^x(-x^3 + 3x) - 3]\,dx$ on a graphing calculator or by looking at the area above and below the x-axis between $x = 0.618$ and $x = 5$. Since there is more area below or by evaluating the definite integral, the conclusion is that $F(5) - F(0.618) < 0$ and that $F(5) < F(0.618)$.

The correct choice is (E).

37. The derivative may be approximated by calculating the slope from any two points near $x = 4$, for example,

$$f'(4) \approx \frac{1.16016 - 1.15782}{4.00100 - 4.00000} = \frac{0.00234}{0.001} = 2.340$$

Using other points gives comparable results.

The correct choice is (D).

38. In order for $f(x)$ to be continuous at $x = 0$,

$$f(0) = \lim_{x \to 0} f(x)$$

Since $\lim_{x \to 0} f(x) = \lim_{x \to 0} \frac{\sin x}{x} = 1$, $f(0) = 1$ and therefore $k = 1$.

The correct choice is (B).

39. Graph the two functions $y = \cos 2x$ and $y = \sin 3x$ in the interval $0 \le x \le 5$. Choices (A) through (D) represent the areas of the regions enclosed by the graphs from left to right. It is easy to see that choice (D) is the largest. Choice (E) has the wrong function "on top" and therefore represents a negative number.

    The correct choice is (D).

40. First graph the function $f(x)$ in the window $[0,1.4]$ by $[-10,10]$. Identify the zero at $x = 1.042$ (which is $\frac{\ln \pi}{\ln 3}$). Then use the derivative function to find the derivative; and, the derivative at the point $(x = 1.042)$ is 3.451. Do not be misled by the vertical asymptote near $x = 0.411$.

    The correct choice is (C).

41. The interval $[0,6]$ is divided into three subintervals $[0,2]$, $[2,4]$, and $[4,6]$. The integral is approximated using the function's value at the midpoint of each interval multiplied by the width of the interval:

    $$(0.25)(2) + (0.68)(2) + (0.95)(2) = 3.76$$

    For additional practice: (A) is the left Riemann sum, (E) is the right Riemann sum and (B) is the trapezoidal approximation, all with three subintervals.

    The correct choice is (D).

42. By the Fundamental Theorem of Calculus (version II),

    $$\begin{aligned}\frac{d}{dx}\int_x^{x^3} \sin(t^2)\,dt &= 3x^2 \sin((x^3)^2) - \sin(x^2)\\ &= 3x^2 \sin(x^6) - \sin(x^2)\end{aligned}$$

    The correct choice is (C).

43. The amount of water, A, after x hours, is given by the accumulation function

    $$A(x) = \int_0^x 300\sqrt{t}\,dt$$

    After 4 hours, there are $\int_0^4 300\sqrt{t}\,dt$ number of gallons.

    $$\int_0^4 300\sqrt{t}\,dt = 300\int_0^4 \sqrt{t}\,dt = 300 \cdot \frac{2}{3} \cdot t^{\frac{3}{2}}\Big|_0^4 = 200\left(4^{\frac{3}{2}} - 0\right) = 1600 \text{ gallons}$$

    The correct choice is (D).

44. Solve $2 \sin x = x$ to obtain $x = 1.89549$. Let $A = 1.89549$.

Using a graphing calculator, evaluate $V = \pi \int_0^A [(2 \sin x)^2 - x^2]\, dx = 6.678$.

The correct choice is (D).

45. $y = xe^x$

$$\frac{dy}{dx} = xe^x + e^x = e^x (x + 1)$$

$$\frac{d^2y}{dx^2} = xe^x + e^x + e^x = 2e^x + xe^x = e^x (x + 2)$$

Continuing this pattern, the nth derivative, $\frac{d^n y}{dx^n} = e^x (x + n)$.

The correct choice is (C).

1a. Graph the function $T(x)$ in a suitable window such as $[0,24]$ by $[50,100]$. Using a graphing calculator, find the value of x which makes $y = 70$. Rounded to the nearest half-hour, $x \approx 8.5$ or at 8:30 am.

1b. Same idea as in part (a). Find the value of x which makes $y = 77$. Rounded to the nearest half-hour, $x \approx 10.5$ or 10:30 am.

1c. Use the cost equation with $x = 8.5$ and $T_0 = 70$:

$$C(x) = 0.16\int_{8.5}^{18} (T(x) - 70)\, dx = 0.16\int_{8.5}^{18}\left[73 - 14\cos\left(\frac{\pi(x-3.4)}{12}\right) - 70\right] dx = \$18.26$$

Use a graphing calculator to evaluate the definite integral.

1d. In a like manner, use $T_0 = 77$ and $x = 10.5$, to first find the cost of cooling at 77°F:

$$C(x) = 0.16\int_{10.5}^{18}\left[73 - 14\cos\left(\frac{\pi(x-3.4)}{12}\right) - 77\right] dx = \$8.79$$

The savings is \$18.26 – \$8.79 = \$9.47 per day.

2a. Solve the equation $\sin(x) = \frac{1}{2}$ to determine where the two graphs intersect. The values are $x = \frac{\pi}{6}$ and $x = \frac{5\pi}{6}$.

$$\text{Area} = \int_{\frac{\pi}{6}}^{\frac{5\pi}{6}} \left(\sin(x) - \frac{1}{2}\right) dx = 0.684 \text{ or } 0.685$$

2b. The thickness of the washer is dx. The outer radius is $\sin(x)$ and the inner radius is $\frac{1}{2}$. Volume $= \pi \int_{\frac{\pi}{6}}^{\frac{5\pi}{6}} \left((\sin(x))^2 - \left(\frac{1}{2}\right)^2\right) dx = 3.005$

2c. The thickness is dx. The diameter of the semi-circles is the length of the chord between the graphs, which is $\sin(x) - \frac{1}{2}$, so the radius of the semi-circles is $\frac{\sin(x) - \frac{1}{2}}{2}$.

$$\text{Volume} = \frac{\pi}{2} \int_{\frac{\pi}{6}}^{\frac{5\pi}{6}} \left(\frac{\sin(x) - \frac{1}{2}}{2}\right)^2 dx = 0.106 \text{ or } 0.107$$

REMINDER: Numeric and algebraic answers need not be simplified, and if simplified incorrectly credit will be lost.

3a. At (5,6) the slope of the tangent line is $h'(5) = 0.7$. The tangent line is

$$\begin{aligned} y(x) &= 0.7(x-5)+6 \\ y(4) &= 0.7(-1)+6 \\ &= 5.3 \text{ inches} \end{aligned}$$

Since $h''(t) > 0$, the function is concave up, so the tangent line lies below the curve and gives an approximation that is less than the true value.

3b.
$$\begin{aligned} V &= (10)(20)\,h \\ \frac{dV}{dt} &= 200\,\frac{dh}{dt} \\ \left.\frac{dV}{dt}\right|_{t=2} &= 200(0.5) = 100 \text{ cubic inches per minute.} \end{aligned}$$

3c.
$$\begin{aligned} h &\approx 0.4(2)+0.5(3)+0.7(4)+1.0(4)+1.1(2) \\ &\approx 11.3 \text{ inches} \end{aligned}$$

$\int_0^{15} h'(t)\,dt$ is the change in the depth of water in the tank, in inches, from $t = 0$ to $t = 15$ minutes.

(Note: $\int_0^{15} h'(t)\,dt$ does *not* represent the depth of water in the tank at $t = 15$ minutes because the depth of water at $t = 0$ minutes is unknown).

3d. The approximation in part (c) is less than the true amount. Since $h''(t) > 0, h'(t)$ is increasing, and the left side values used in the approximation are all less than the true values.
(Note: For a function which is increasing and concave up, the rectangles in a left Riemann sum all lie below the curve.)

4a. $y = \frac{x}{x+C}, x \neq C, C \neq 0$ and from the given differential equation: $y' = \frac{y - y^2}{x}$

$$y' = \frac{(x+C)(1) - x(1)}{(x+C)^2}$$

$$y' = \frac{C}{(x+C)^2}$$

$$= \frac{\frac{x}{x+C} - \frac{x^2}{(x+C)^2}}{x}$$

$$= \frac{x(x+C) - x^2}{x(x+C)^2}$$

$$= \frac{C}{(x+C)^2}$$

By substituting (0,0) into $y = \frac{x}{x+C}: 0 = \frac{0}{0+C}$ is true, so regardless of the value of C, (0,0) is on the solution curve.

4b. Substituting (1,2) into $y = \frac{x}{x+C}$ gives $2 = \frac{1}{1+C}$ and $C = -\frac{1}{2}$. The particular solution is $y = \frac{x}{x - \frac{1}{2}}$, and the slope at (0,0) is $y'(0) = \frac{-\frac{1}{2}}{\left(0 - \frac{1}{2}\right)^2} = -2$

4c. The asymptotes are $x = \frac{1}{2}$ and $y = 1$

4d.

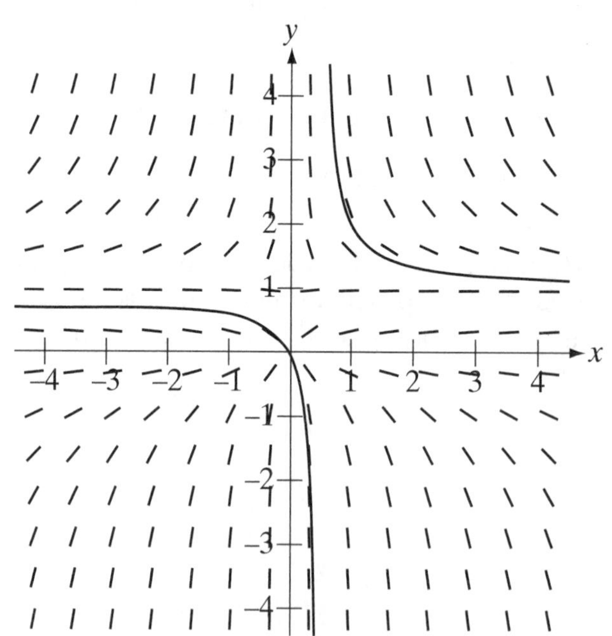

5a. Using implicit differentiation:

$$(\sec^2 y)\left(\frac{dy}{dx}\right) = 1 + \frac{dy}{dx}$$

Solving for $\frac{dy}{dx}$, $\frac{dy}{dx} = \frac{1}{\sec^2 y - 1}$ or $\frac{1}{\tan^2 y}$ or $\cot^2 y$.

5b. The tangents are vertical when the denominator of the derivative is zero.

Solving $\tan^2 y = 0$ in the given interval yields $y = -\pi, 0,$ and π.

Then substitute to find the corresponding values of x:

$(-\pi, \pi)$, $(0,0)$, and $(\pi, -\pi)$.

5c. From part (a), $y' = \frac{1}{\tan^2 y} = \tan^{-2} y = \cot^2 y$

Using the Chain Rule, $\frac{d^2y}{dx^2} = -2(\tan y)^{-3}(\sec y)^2 \frac{dy}{dx}$

$$= -2(\tan y)^{-3}(\sec y)^2(\tan y)^{-2}$$

$$= -2\frac{\sec^2 y}{\tan^5 y}$$

$$= -2\sec^2 y \cot^5 y$$

Any one of the last three solutions is acceptable.

Alternately from part (a), $y' = \cot^2 y$ could have been used to find $\frac{d^2y}{dx^2}$.

6a. f has a relative maximum at $x = 3$. At $x = 3$, the derivative $f'(x)$ changes from positive to negative indicating the the function f changes from increasing to decreasing at $x = 3$. (First Derivative Test)

6b. f has a relative minimum at $x = -3$. At $x = -3$, the derivative $f'(x)$ changes from negative to positive indicating that the function f changes from decreasing to increasing at $x = -3$. (First Derivative Test)

6c. The function f is concave down when $f'' < 0$. The second derivative $f''(x) < 0$ when the first derivative is decreasing. Looking at the given graph of $f'(x)$, this happens in the intervals $-2 \le x \le 0$ and $2 \le x \le 4$.

6d. Since the derivative $f(x)$ is an even function and $f(0) = 0$, the function $f(x)$ must be an odd function. Therefore, $f(x)$ is symmetric to the origin, and $\int_{-a}^{a} f(x)\,dx = 0$.

Sample Examination III

1. The area of the region is represented by $\int_1^3 (3x^2 + 2x)\,dx$.

   $$\int_1^3 (3x^2 + 2x)\,dx = x^3 + x^2\Big|_1^3 = (27 + 9) - (1 + 1) = 36 - 2 = 34$$

   The correct choice is (B).

2. Rewrite the function without the absolute value signs:

   $$f(x) = \begin{cases} -x, & x < 0 \\ 0, & x = 0 \\ x, & x > 0 \end{cases}$$

   $$\begin{aligned} \text{Then } \int_{-4}^{2} f(x)\,dx &= \int_{-4}^{0} (-x)\,dx + \int_{-4}^{0} x dx \\ &= -\frac{x^2}{2}\Big|_4^0 + \frac{x^2}{2}\Big|_0^2 \\ &= 0(0 - (-8)) + (2 - 0) \\ &= 10 \end{aligned}$$

   The correct choice is (E).

3. The units of a definite integral or a Riemann sum are the units of the dependent variable (the integrand) multiplied by the units of the independent variable, *ds*, in this case

   $(gallons/mile) \times (miles/hour) = gallons/hour$

   The correct choice is (C).

4. <u>Given</u>: $\frac{dr}{dt} = 2$

   <u>Find</u>: $\frac{dV}{dt}$ when $r = 10$

   Since the volume of a sphere is $V = \frac{4}{3}\pi r^3$, differentiate both sides of the equation with respect to t.

$$\frac{dV}{dt} = 4\pi r^2 \frac{dr}{dt}$$

$$\text{or } \frac{dV}{dt} = 4\pi(10)^2(2) = 800\pi$$

   The correct choice is (D).

5. Since the velocity is positive over the given interval, the distance the particle travels from $t = 1$ to $t = 3$ can be represented by

$$\int_1^3 v(t)\,dt = \int_1^3 t^2\,dt = \frac{t^3}{3}\Big|_1^3 = 9 - \frac{1}{3} = \frac{26}{3}$$

   The correct choice is (B).

6. By the Fundamental Theorem of Calculus: $\int_1^4 f'(x)\,dx = f(4) - f(1) = 2 - (-5) = 7$

   The correct choice is (D).

7. The choices give three different initial conditions for the unknown differential equation. For each trace along the slope field and see where the curves go. Starting above (0,2) the path moves down to the right leveling off near $y = 2$. So I is true. Starting between 0 and 2 on the y-axis and moving right the curve rises towards $y = 2$, so II is true. Starting on the y-axis below the origin and tracing to the right leads away from the x-axis, so III is false.

   The correct choice is (D).

8. Let $u = x^2 + 1$

$$du = 2x\,dx \Rightarrow x\,dx = \frac{1}{2}du$$

$$\int \frac{x}{x^2+1}dx = \frac{1}{2}\int \frac{du}{u} = \frac{1}{2}\ln|u| = \frac{1}{2}\ln(x^2+1)\Big|_1^3 = \frac{1}{2}(\ln 10 - \ln 2)$$

Since $\ln 10 - \ln 2 = \ln\left(\frac{10}{2}\right) = \ln 5$, therefore, $\int_1^3 \frac{x}{x^2+1}dx = \frac{1}{2}\ln 5$.

The correct choice is (D).

9. Using the properties of logarithms and the Chain Rule:

$$\frac{d}{dx}\ln\left(\frac{1}{x^2-1}\right) = \frac{d}{dx}(\ln 1 - \ln(x^2-1))$$
$$= \frac{d}{dx}(-\ln(x^2-1))$$
$$= -\frac{2x}{x^2-1} = \frac{2x}{1-x^2}$$

The correct choice is (B).

10. To find an antiderivative of $2\tan x$, integrate $2\tan x$,

$$\int 2\tan x\,dx = 2\int \tan x\,dx = 2\ln|\sec x| = 2\ln(\sec x), \text{ since } 0 \le x < \frac{\pi}{2}.$$

Since $b\ln a = \ln a^b$, $2\ln(\sec x) = \ln(\sec^2 x)$.

The correct choice is (C).

11. Since the function is concave down, the first derivative must always decrease.

In table (A): $\frac{f(5)-f(4)}{5-4} = -4$ so by the Mean Value Theorem there must be a value, c_1, between $x = 4$ and $x = 5$ where $f'(c_1) = -4$. Also $\frac{f(6)-f(5)}{6-5} = -4$ so there must be a value c_2, between $x = 5$ and $x = 6$ where $f'(c_2) = -4$. Thus, the derivative goes from -4 to -3 (given) to -4. Since the derivative is not decreasing throughout the interval, (A) cannot be the correct table.

Doing a similar analysis on the other tables, the derivatives, in the same order, are

(B) –2, –3, –4

(C) –2, –3, –1

(D) –2, –3, –2

(E) –5, –3, –1

This eliminates (C), (D) and (E). The remaining choice (B) is the only one that *could* be the table of values for f.

Although what happens between the values given in the table is unknown, the slopes in (B) clearly indicate a decrease from –2 to –4 and therefore (B) is the <u>best</u> possible choice. In all of the other four choices it is clear that the slopes do not continually decrease.

The correct choice is (B).

12. $y' = \dfrac{k(k+x)-(kx+8)}{(k+x)^2}$, or

$$y' = \frac{k^2-8}{(k+x)^2}$$

Since $y = x+4$ is the line tangent to y at $x = -2$, its slope is $y' = 1$.

By substituting $x = -2$ and $y'(-2) = 1$,

$$1 = \frac{k^2-8}{(k-2)^2} \Rightarrow k^2-8 = (k-2)^2 \text{ or } k^2-8 = k^2-4k+4, \text{and } k = 3.$$

The correct choice is (D).

13. Notice that the subintervals are not of equal width. Multiplying the function on the right side of each subinterval times the width of that subinterval gives

$$7(1)+11(3)+12(2)+8(1) = 72$$

The correct choice is (E).

14. Since the function has a derivative for all $x \neq 2$, it is continuous everywhere else. The derivative is positive indicating that the function is always increasing. The function must increase to infinity on the left side of the asymptote and then increase *from* $-\infty$ on the right side. In other words, this is an odd vertical asymptote. Thus the limit in II should equal $-\infty$ and the limit in I does not exist since the two one-sided limits are different. Only <u>III is true</u>. An example of such a function is $f(x) = \frac{-1}{x-2}$; its graph is shown below:

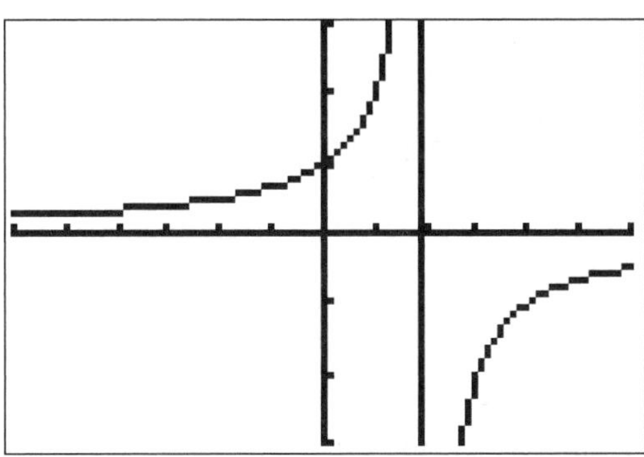

The correct choice is (C).

15. By the First Derivative Test, for a function to have a minimum value the derivative must change sign from negative to positive, this is choice (D). Choices (A) and (C) do not have enough information to justify a maximum or minimum. Choice (B) is wrong because for a maximum, the derivative must change from positive to negative. For Choice (E) to be a correct justification by the Second Derivative Test, you also need to know that $f'(3) = 0$.

The correct choice is (D).

16. At a point of inflection where the graph changes from concave up to concave down the *second* derivative ($g'' = f'$) changes from positive to negative. This occurs where the *first* derivative ($g' = f$) changes from increasing to decreasing. The value is $x = 5$. Since f has a relative maximum at $x = 5$, f' changes from positive to negative and thus g changes from concave up to concave down.

The correct choice is (D).

17. First, observe that $f(x)$ is increasing where $f'(x) \geq 0$ is decreasing where $f'(x) < 0$. In the given graph of $f'(x)$, $f'(x) \geq 0$ for $x \geq -2$; $f'(x) < 0$ for $x < -2$. Therefore, $f(x)$ is decreasing for $x < -2$ and increasing for all $x \geq -2$. From the given choices, the only graph that is decreasing to the left of $x = -2$ and increasing to the right of $x = -2$ is (A).

The correct choice is (A).

18. By the Fundamental Theorem of Calculus (version II),

$$\frac{d}{dx}\int_x^0 \frac{du}{1+u^2} = \frac{d}{dx}\left(-\int_0^x \frac{du}{1+u^2}\right) = \frac{-1}{1+x^2}$$

The correct choice is (B).

19. The segment with endpoints $(-2,26)$ and $(10,2)$ has a slope of -2. Therefore, by the Mean Value Theorem, the line through $(4,23)$ will be parallel to the segment and tangent to the graph; its equation is $y - 23 = -2(x - 4)$. Its y-intercept is 31.

The correct choice is (C).

20. Since at each point (x,y) on a certain curve, the slope of the curve is $4xy$, $\frac{dy}{dx} = 4xy$.
To find y (the equation of the curve), solve the following separable differential equation.

$$\frac{dy}{dx} = 4xy \Rightarrow \frac{dy}{y} = 4x\,dx \Rightarrow \int \frac{dy}{y} = \int 4x\,dx$$

$$\Rightarrow \ln|y| = 2x^2 + C_1$$

$$\Rightarrow y = Ce^{2x^2}$$

Since the curve contains the point $(0,4)$, $4 = Ce^0 \Rightarrow C = 4$.

Therefore, $y = 4e^{2x^2}$.

The correct choice is (C).

21. To find the point of inflection, set $y'' = 0$ and check the concavity near that point (to make sure it switches from concave up to concave down or vice versa).

$y' = 3x^2 - 12x$

$y'' = 6x - 12$

$y'' = 0$ when $6x - 12 = 0$ or $x = 2$

	$x < 2$	$x > 2$
f''	$-$	$+$

Therefore, only (2, –16) is a point of inflection.

The line tangent to the curve $y = x^3 - 6x^2$ has a slope of $y' = 3x^2 - 12x$; and when $x = 2$, $y' = -12$.

An equation of the tangent line is $y - y_1 = m(x - x_1)$

$$y - (-16) = -12(x - 2)$$

$$y + 16 = -12x + 24$$

$$y = -12x + 8$$

The correct choice is (A).

22. Let $u = 2x$, $du = 2\,dx$ and $dx = \frac{1}{2}du$

$$\int \cos 2x\,dx = \frac{1}{2}\int \cos u\,du = \frac{1}{2}\sin 2x\Big|_0^k = \frac{1}{2}\sin 2k$$

Since $\frac{1}{2}\sin 2k = \frac{1}{2}$, $\sin 2k = 1 \Rightarrow 2k = \frac{\pi}{2}$ and $k = \frac{\pi}{4}$

The correct choice is (B).

23. The only properties which are true are I and III.

Property II is false because only a constant may be "taken out" of the integrand.

The correct choice is (C).

24. $f(x) = \sqrt{e^{2x} + 1} = (e^{2x} + 1)^{\frac{1}{2}}$ and $f'(x) = \frac{1}{2}(e^{2x} + 1)^{-\frac{1}{2}} \cdot 2e^{2x} = \dfrac{e^{2x}}{\sqrt{e^{2x} + 1}}$

Therefore, $f'(0) = \dfrac{e^0}{\sqrt{e^0 + 1}} = \dfrac{1}{\sqrt{2}} = \dfrac{\sqrt{2}}{2}$

The correct choice is (C).

25. Method 1:

Notice that $x^2y + yx^2 = 2x^2y$.

So $2x^2y = 6 \Rightarrow y = \frac{6}{2x^2} = \frac{3}{x^2} = 3x^{-2}$

To find $\frac{d^2y}{dx^2}$, take the second derivative:

$$y = 3x^{-2}$$

$$y' = -6x^{-3}$$

$$y'' = \frac{d^2y}{dx^2} = 18x^{-4} \text{ or } \frac{18}{x^4}$$

So $\frac{d^2y}{dx^2}$ at $x = 1$ is $\frac{18}{1^4} = 18$

For this problem there was really no need for implicit differentiation.

Method II:

Using implicit differentiation, first find $\frac{dy}{dx}$ (the first derivative):

$$\left(x^2\frac{dy}{dx} + 2xy\right) + \left(2xy + x^2\frac{dy}{dx}\right) = 0 \Rightarrow 4xy + 2x^2\frac{dy}{dx} = 0 \Rightarrow \frac{dy}{dx} = \frac{-4xy}{2x^2} = \frac{-2y}{x}$$

Now, since $\frac{dy}{dx} = \frac{-2y}{x}$, find $\frac{d^2y}{dx^2}$ by differentiating $\frac{-2y}{x}$ implicitly.

$$\frac{d^2y}{dx^2} = \frac{x\left(-2\frac{dy}{dx}\right) - (-2y)}{x^2} = \frac{-2x\left(\frac{-2y}{x}\right) + 2y}{x^2} = \frac{6y}{x^2}$$

Note: $\frac{-2y}{x}$ was substituted for $\frac{dy}{dx}$ in the equation above.

Therefore $\frac{d^2y}{dx^2} = \frac{6y}{x^2}$ and at (1,3), $\frac{d^2y}{dx^2} = \frac{6(3)}{1^2} = 18$

The correct choice is (E).

26. Since every cross section is a semicircle, Area $= \frac{1}{2}\pi(\text{radius})^2 = \frac{1}{2}\pi\left(\frac{1}{2}y\right)^2 = \frac{\pi}{8}y^2$.

$$\text{Volume} = \int_0^4 \frac{\pi}{8}\left(\frac{4-x}{2}\right)^2 dx \text{ (since } x + 2y = 4 \Rightarrow 2y = 4 - x \text{ or } y = \tfrac{4-x}{2}\text{)}$$
$$= \frac{\pi}{32}\int_0^4 (4-x)^2 dx$$
$$= \frac{\pi}{32}\int_0^4 (16 - 8x + x^2)\, dx$$
$$= \frac{\pi}{32}\left(16x - 4x^2 + \frac{x^3}{3}\right)\Big|_0^4$$
$$= \frac{\pi}{32}\left[\left(64 - 64 + \frac{64}{3}\right) - (0)\right]$$
$$= \frac{\pi}{32}\left(\frac{64}{3}\right) = \frac{2\pi}{3}$$

The correct choice is (A).

27. The velocity of a particle, $v(t)$, is increasing when $v'(t) > 0$.

$$x(t) = (t+1)(t-3)^3$$
$$v(t) = x'(t) = (t+1)\cdot 3(t-3)^2 + (t-3)^3 = (t-3)^2\,[3(t+1) + (t-3)] = (t-3)^2\,(4t)$$
$$v'(t) = 4(t-3)^2 + (4t)\cdot 2(t-3) = 4(t-3)^2 + (8t)(t-3) = 4(t-3)[(t-3) + 2t]$$
$$= 4(t-3)(3t-3)$$

$v'(t) = 0$ when $t = 1$ and $t = 3$.

	$t < 1$	$1 < t < 3$	$t > 3$
$v'(t)$	+	–	+

Therefore, $v'(t) > 0$ when $t < 1$ or $t > 3$.

The correct choice is (D).

28. The general solution to the differential equation $y' = ky$ is $y = Ce^{kt}$ where C is the initial population. It is given that the initial population, $C = 1{,}500$ and this population is 6,000 after the first 2 days.

$$\begin{aligned} 6{,}000 &= 1{,}500e^{2k} \\ 4 &= e^{2k} \\ \ln 4 &= 2k \\ k &= \frac{\ln 4}{2} = \frac{2\ln 2}{2} = \ln 2 \\ \text{Thus } y &= 1{,}500e^{t\ln 2} \\ \text{At the time } t=3, y &= 1{,}500e^{3\ln 2} \\ y &= 1{,}500e^{\ln(2^3)} \\ y &= 1{,}500\,(2^3) \\ y &= 1{,}500\,(8) \end{aligned}$$

Therefore, the population will have increased by a factor of 8.

The correct choice is (D).

29. The average value of $f(x)$ is $\frac{1}{7-3}\int_3^7 f(x)\,dx = 12$

$$\Rightarrow \frac{1}{4}\int_3^7 f(x)\,dx = 12$$

$$\Rightarrow \int_3^7 f(x)\,dx = 48$$

The correct choice is (E).

30. The volume of the solid generated is given by $V = \pi\int_0^{\pi} [\cos(\cos x)]^2 dx = 6.04$.

The correct choice is (C).

31. By the Quotient Rule $h'(x) = \dfrac{g(x)f'(x) - f(x)g'(x)}{[g(x)]^2}$.

Then $h'(3) = \dfrac{g(3)f'(3) - f(3)g'(3)}{[g(3)]^2} = \dfrac{(3)(-\frac{1}{3}) - (1)(1)}{3^2} = -\dfrac{2}{9}$

The values of the functions are read from the graph; the values of the derivatives are the slopes of the lines at $x = 3$.

The correct choice is (A).

32. Since the area of the region for $0 \le x \le c$ is equal to the area of the region for $c \le x \le \frac{\pi}{2}$,

$$\int_0^c \cos x\, dx = \int_c^{\frac{\pi}{2}} \cos x\, dx$$

$$\sin x\Big|_0^c = \sin x\Big|_c^{\frac{\pi}{2}}$$

$$\sin c = 1 - \sin c$$

$$2 \sin c = 1$$

$$\sin c = \tfrac{1}{2} \Rightarrow c = \tfrac{\pi}{6}$$

The correct choice is (B).

33. If the derivative is negative the function is decreasing.

$$\begin{aligned} f'(x) &= -3x^2 + 12x + 15 \\ &= -3(x+1)(x-5) \end{aligned}$$

The derivative is negative when $x < -1$ or $x > 5$, so this is where the function decreases.

The correct choice is (D).

34. $$\begin{aligned} f(1) - f(0) &= \int_0^1 \frac{\tan^2 x}{x^2+1}\,dx \\ \frac{1}{2} - f(0) &\approx 0.3446 \\ -f(0) &\approx 0.3446 - 0.500 \\ f(0) &\approx 0.1554 \end{aligned}$$

There is no easy way to find the antiderivative for the given expression. Evaluate the definite integral on your calculator and use the Fundamental Theorem of Calculus.

The correct choice is (B).

35. Graph the second derivative. The function $f(x)$ will have a point of inflection when the second derivative changes sign. The second derivative changes sign (crosses the x-axis) six times in this interval.

The correct choice is (C).

36. In order for the local linear approximation to be greater than or equal to the function's value on an interval, the tangent line must lie entirely above the curve (except at the point of tangency). This means that the curve must be concave downward in the interval and the second derivative will therefore be negative.

    The correct choice is (C).

37. The function increases most rapidly when it has the largest positive slope or when its derivative is largest. $f'(x) = \frac{5.8\pi}{4}\cos\left(\frac{\pi x}{4}\right) + \frac{15.7\pi}{3}\sin\left(\frac{\pi x}{3}\right)$. Graph the derivative in a suitable window such as $x[0,12]$ by $y[-21,21]$. (See figure).

    The maximum is the highest point on this graph, which is near $x = 7.566$.

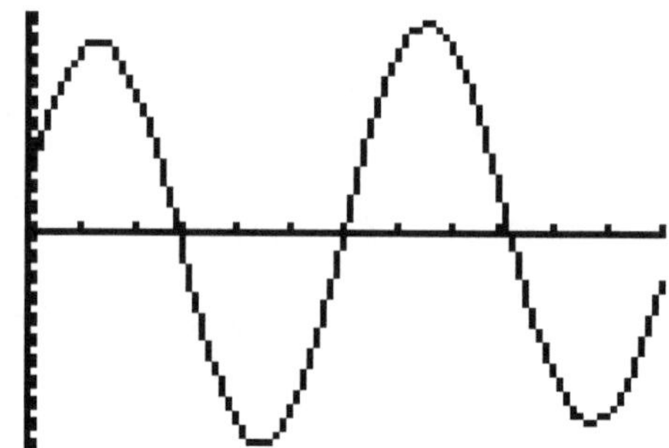

    The correct choice is (D).

38. The given limit is the derivative of $g(x)$ at $x = 3$. Since $g'(3)$ is negative, the function must be decreasing at $x = 3$.

    The correct choice is (A).

39. Begin by graphing y', the function's derivative, in the window $[0,1]$ by $[-2,2]$. Then change XMAX to 0.1, then 0.01, then 0.001, etc. Each time you will see a different graph! As x decreases from 1 to zero, $\ln x$ decreases from zero to $-\infty$. The $\sin(\ln x)$ acts like $\sin x$ as x goes from zero to $-\infty$, only it does it in the space from one to zero (like a compressed spring). The zeros of the function occur when $\sin(\ln x) = 0$, which is when $\ln x = \pm k\pi$, $k = 0,1,2,3,\ldots$, or when $x = e^{\pm k\pi}$, $k = 1,2,3,4,\ldots$. All of the negative powers are in the given interval, therefore there are an infinite number of zeros. HOWEVER, since each zero is over an order of magnitude

larger than the next smaller zero, it is impossible to "see" more than three roots in any graphing window. Be aware of the limitations of the graphing calculator. Therefore, there are more than 4 relative extreme values for this function.

The correct choice is (E).

40. If $f^{-1}(x) = g(x)$, then $g'(x) = \dfrac{1}{f'(g(x))}$ and $g'(0) = \dfrac{1}{f'(g(0))}$.

Solve on a graphing calculator $x^3 - 7x^2 + 25x - 39 = 0$. Thus $x = 3$.

Since $f(3) = 0$, $g(0) = 3$.

$f'(x) = 3x^2 - 14x + 25$, $f'(3) = 3(3^2) - 14(3) + 25 = 10$

$g'(0) = \dfrac{1}{f'(3)} = \dfrac{1}{10}$

The correct choice is (C).

41. The units of $4\int_0^2 \rho(x)\,dx$ is given as *people*. The units of 4 and x are miles. Therefore

$$\text{miles} \cdot (\text{units of } \rho(x)) \cdot \text{miles} = \text{people}$$

$$\text{units of } \rho(x) = \frac{\text{people}}{\text{miles} \cdot \text{miles}} = \text{people per square mile}$$

The correct choice is (E).

42. The area A of the rectangle may be represented by $A = 2x\sqrt{64 - x^2}$ where $2x$ is the base of the rectangle and its height is $y = \sqrt{64 - x^2}$.

$$A' = (2x)\left(\frac{-x}{\sqrt{64 - x^2}}\right) + 2\sqrt{64 - x^2} = \frac{-2x^2 + 2(64 - x^2)}{\sqrt{64 - x^2}} = \frac{128 - 4x^2}{\sqrt{64 - x^2}}$$

Reject the critical values of $x = \pm 8$ from the denominator, since the rectangle's height would equal zero.

The only critical values to consider are when $128 - 4x^2 = 0 \Rightarrow x = \pm\sqrt{32}$.

Reject $x = -\sqrt{32}$ for a length and choose $x = \sqrt{32}$ as the maximum. Since A' switches from + to – at $x = \sqrt{32}$.

Therefore, the maximum area is $A = 2(\sqrt{32})(\sqrt{32}) = 64$.

The correct choice is (E).

43. If f is differentiable at $x = 0$, then it is continuous at $x = 0$.

If f is to be continuous at $x = 0$, then $f(0) = \lim_{x\to 0} f(x)$.

$f(0) = b$; $\lim_{x\to 0^+} f(x) = b$ and $\lim_{x\to 0^-} f(x) = 3 \Rightarrow b = 3$.

$$f'(x) = \begin{cases} -e^{-x}, & \text{for } x < 0 \\ a, & \text{for } x \geq 0 \end{cases}$$

If f is differentiable at $x = 0$, then $\lim_{x\to 0^+} f'(x) = \lim_{x\to 0^-} f'(x)$.

$\lim_{x\to 0^+} f'(x) = a$ and $\lim_{x\to 0^-} f'(x) = -1 \Rightarrow a = -1$.

Therefore $a = -1$ and $b = 3$, or $a + b = 2$.

The correct choice is (D).

44. Using a calculator the amount is found by integrating $R(t)$: $\int_2^{10} 100\left(\frac{5t^2 - t - 1}{5t^2 + t}\right) dt \approx 731$ gallons.

The correct choice is (C).

45. The function f is a parabola with a maximum point at (2,2). When $b = 2$ the horizontal line $y = 2$ is tangent at (2,2). As b increases from 2, the tangent lines will move to the right of the maximum point. The "last" line tangent in the first quadrant will be just before the zero of f at $x = 3$. Find the equation of the tangent line at $x = 3$. The slope is $f'(x) = -4(x - 2)$ and $f'(3) = -4$. The tangent line at this point is $y = -4(x - 3)$ or $y = -4x + 12$. Thus b must be between 2 and 12. (Note that $b \neq 12$ on the technicality that the axis is not in the first quadrant.)

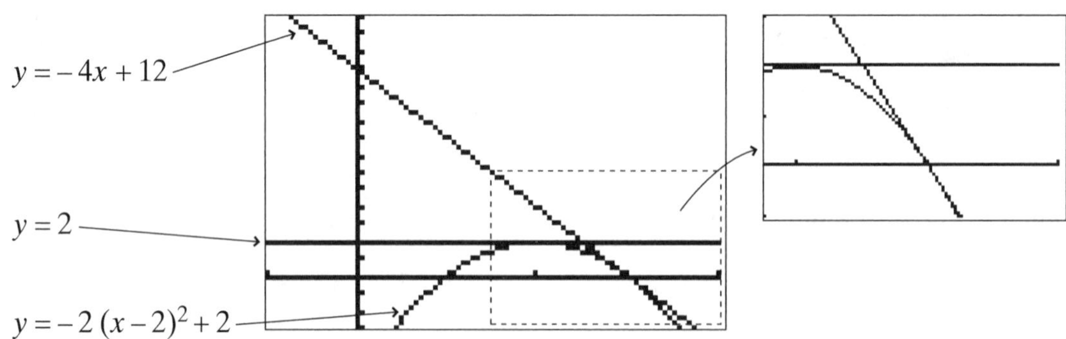

The correct choice is (C).

1a. $v(3) = -0.841$; speed $= |v(3)| = 0.841$; acceleration $= v'(3) = 3.906$.

1b.

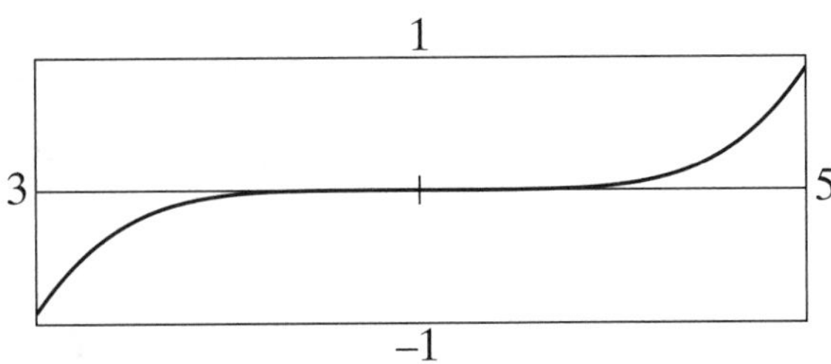

1c. The velocity is increasing on the interval [3,5]. Its graph is rising throughout the interval.

(Note: The fact that $v(4) = 0$ and $a(4) = 0$ does not matter. On the velocity graph every point, including (4,0) is above those to its left and below those to its right. Therefore the velocity is increasing in the interval.)

1d. Distance $= \int_3^5 |v(t)|\, dt = 0.293$

2a. $\int_0^9 \frac{5000}{x^3+50}\,dx \approx 415.420$ or 415.421 cubic meters

2b. $S(6) - R(6) = \frac{5000}{6^3+50} - 23.9665\sqrt{6} = -39.908$ or -39.909 cubic meters per hour. This means that the amount of sand in the bin is decreasing at the rate of 39.908 (or 39.909) cubic meters per hour when $t = 6$.

2c. The amount of sand, $A(t)$, in the bin is given by $A(t) = \int_0^t S(x)\,dx - \int_1^t R(x)\,dx$. The maximum occurs when $A'(t) = S(t) - R(t) = 0$ or when $S(t) = R(t)$.

OR: Before the time when $S(t) = R(t)$ the rate at which sand is being poured into the bin is greater than the rate at which it is being taken out of the bin. After this point the rate at which it is taken out is greater than the rate at which it is pouring into the bin. Therefore the maximum must occur when $S(t) = R(t)$.

2d. From part (a) the amount of sand put into the bin in the course of the day is about 415.42091 cubic meters. The amount that is taken out is $\int_1^9 23.9665\sqrt{t}\,dt \approx 415.41933$ cubic meters. Therefore, at the end of the day, the bin has 0.00158 or 0.002 cubic meters of sand.

<u>REMINDER</u>: Numerical and algebraic answers need not be simplified, and if simplified incorrectly credit will be lost.

3a. $\text{Area} = \int_0^4 \sqrt{x} - \frac{1}{2}x\,dx = \frac{2}{3}x^{\frac{3}{2}} - \frac{x^2}{4}\Big|_0^4$

$$= \frac{2}{3}(4)^{\frac{3}{2}} - \frac{4^2}{4} - 0$$

$$= \frac{4}{3}$$

3b. The circular cross sections will have a thickness of dy. The outer radius is $x = 2y$ and the inner radius is $x = y^2$. Integrate between the y values of 0 and 2.

$\text{Volume} = \pi\int_0^2 \left((2y)^2 - (y^2)^2\right) dy = \pi\left(\frac{4}{3}y^3 - \frac{y^5}{5}\right)\Big|_0^2$

$$= \pi\left(\frac{4}{3}(2^3) - \frac{2^5}{5} - 0\right)$$

$$= \frac{64\pi}{15}$$

3c. The outer radius is $\sqrt{x} - (-3)$ and the inner radius is $\frac{x}{2} - (-3)$. The thickness is dx.

$\text{Volume} = \pi\int_0^4 \left((\sqrt{x} - (-3))^2 - \left(\frac{x}{2} - (-3)\right)^2\right) dx$

4a. At the points of tangency, the slope of f is $2a$ and the slope of g is $2(b-3)$. Both of these are the slope of the same tangent line so $2a = 2(b-3)$ or $a = b - 3$.

4b. Using the two points of tangency, the slope of T is $\dfrac{(b-3)^2 - (a^2-3)}{b-a}$. Substituting $b - 3 = a$ and expanding gives

$$\frac{a^2 - (a^2 - 3)}{a + 3 - a} = \frac{3}{3} = 1$$

So the slope of T is 1 and

$$2a = 1 \text{ and } 2(b-3) = 1, \text{or}$$

$$a = \frac{1}{2} \text{ and } b = \frac{7}{2}$$

4c. Using the point $(a, a^2 - 3)$ when $a = \frac{1}{2}$, $y - \left(\left(\frac{1}{2}\right)^2 - 3\right) = 1\left(x - \frac{1}{2}\right)$ or $y = x - \frac{13}{4}$

4d. $\int_{\frac{1}{2}}^{2} \left(x^2 - 3 - \left(x - \frac{13}{4}\right)\right) dx + \int_{2}^{\frac{7}{2}} \left((x-3)^2 - \left(x - \frac{13}{4}\right)\right) dx$

5a. To find where the function is increasing, analyze the derivative:

$$\begin{aligned} f'(x) &= 12 + 6x - 6x^2 \\ &= -6(x^2 - x - 2) \\ &= -6(x-2)(x+1) \end{aligned}$$

$$f'(x) \quad \xleftarrow{\quad - \quad} \underset{-1}{\overset{0}{|}} \xrightarrow{\quad + \quad} \underset{2}{\overset{0}{|}} \xrightarrow{\quad - \quad}$$

The derivative is positive between –1 and 2, so the function increases on the interval $[-1,2]$.

5b. From (a), the relative maximum occurs when $x = 2$.

$$\begin{aligned} f(2) &= k + 12(2) + 3(2)^2 - 2(2)^3 \\ 4 &= k + 24 + 12 - 16 \\ -16 &= k \end{aligned}$$

5c. $f''(x) = 6 - 12x$, $f''(x) = 0$ when $x = \frac{1}{2}$, and $f''(x) > 0$ when $x < \frac{1}{2}$. So the function is concave upwards on the interval $(-\infty, \frac{1}{2})$.

5d. From (a) the relative minimum occurs at $x = -1$

$$\begin{aligned} f(-1) &= -16 + 12(-1) + 3(-1)^2 - 2(-1)^3 \\ &= -23 \end{aligned}$$

The relative minimum value is -23.

6a. $g(4.5) = 13.5$. This is the area under the graph of f in the first quadrant between $x = 0$ and $x = 4.5$.

$g'(4.5) = f(4.5) = 0$; $g''(4.5) = f'(4.5) = -2$, the slope of the line segment containing (4.5,0).

6b. Average value of f equals $\frac{1}{5-(-3)}\int_{-3}^{5} f(x)\,dx = \frac{1}{8}\left[\int_{-3}^{3} f(x)\,dx + \int_{3}^{5} f(x)\,dx\right]$

$$= \frac{1}{8}[27+2] = \frac{29}{8}$$

The two integrals are found by finding the signed areas under the graph of f.

6c. There is a point of inflection where f, the derivative of g, has a maximum or minimum. This occurs only at $x = 5$, where f changes from decreasing to increasing, or where f' changes from negative to positive.

6d. The function g increases from $x = -3$ to $x = 4.5$, then decreases until $x = 7$, and then increases again until $x = 9$. There is a maximum at (4.5,13.5) (from part (a)). There is an endpoint maximum at $x = 9$. Use the signed area to find $g(9)$:

$$\begin{aligned} g(9) &= g(4.5) + \int_{4.5}^{9} f(x)\,dx \\ &= 13.5 - 0.25 \end{aligned}$$

The endpoint maximum occurs at (9, 13.25).

Sample Examination IV

1. The limit may be found by rationalizing the denominator of the fraction:

$$\lim_{x\to b}\frac{b-x}{\sqrt{x}-\sqrt{b}}=\lim_{x\to b}\frac{b-x}{\sqrt{x}-\sqrt{b}}\cdot\frac{\sqrt{x}+\sqrt{b}}{\sqrt{x}+\sqrt{b}}=\lim_{x\to b}\frac{(b-x)(\sqrt{x}+\sqrt{b})}{(x-b)}$$

$$=\lim_{x\to b}-(\sqrt{x}+\sqrt{b})=-2\sqrt{b}$$

The correct choice is (A).

2. The slope of the tangent line is the value of the derivative at $x = 0$.

$$f'(x) = -\sin x + 2\sec^2(2x)$$
$$f'(0) = -0 + 2(1) = 2$$

So the tangent line is

$$y - 1 = 2(x - 0)$$

$$y = 2x + 1$$

The correct choice is (B).

3. Let's examine each of the choices:

(A) $\lim_{x\to 3} f(x) = 1$ is <u>not</u> true because $\lim_{x\to 3^+} f(x) = \lim_{x\to 3^-} f(x) = \lim_{x\to 3} f(x) = 3$.
It is true that $f(3) = 1$, but $\lim_{x\to 3} f(x) = 3$.

(B) $\lim_{x\to 4} f(x) = 3$ is <u>not</u> true because $\lim_{x\to 4^+} f(x) = 3$ and $\lim_{x\to 4^-} f(x) = 1\frac{1}{2}$.
For that matter, $\lim_{x\to 4} f(x)$ does not exist since $\lim_{x\to 4^+} f(x) \neq \lim_{x\to 4^-} f(x)$.

(C) Obviously $f(x)$ is <u>not</u> continuous at $x = 3$, because from choice (A) $f(3) = 1$ but $\lim_{x\to 3} f(x) = 3$.

(D) $f(x)$ is continuous at $x = 5$ <u>is true</u> since $f(5) = 3$ and $\lim_{x\to 5} f(x) = 3$.

(E) $\lim_{x\to 6} f(x) = f(6)$ is <u>not</u> true because $\lim_{x\to 6} f(x)$ is undefined and $f(6) = 1$.
$\lim_{x\to 6} f(x)$ is undefined because $\lim_{x\to 6^+} f(x) \neq \lim_{x\to 6^-} f(x)$, i.e. $1 \neq 4$.

The correct choice is (D).

4. To find $f'(2)$, investiage $\lim_{h \to 0} \frac{f(2+h)-f(2)}{h}$.

$$\lim_{h \to 0^+} \frac{f(2+h)-f(2)}{h} = \lim_{h \to 0^+} \frac{(3+|h|)-3}{h} = \lim_{h \to 0^+} \frac{|h|}{h} = \lim_{h \to 0^+} \frac{h}{h} = 1$$

$$\lim_{h \to 0^-} \frac{f(2+h)-f(2)}{h} = \lim_{h \to 0^-} \frac{(3+|h|)-3}{h} = \lim_{h \to 0^-} \frac{|h|}{h} = \lim_{h \to 0^-} \frac{-h}{h} = -1$$

Since the right and left limits are different, $\lim_{h \to 0} \frac{f(2+h)-f(2)}{h}$ does not exist, that is, $f'(2)$ is undefined.

The correct choice is (E).

5. Let $u = 3x + 5$

$du = 3\,dx$ or $dx = \frac{1}{3}du$

$$\int (3x+5)^2 dx = \frac{1}{3}\int u^2 du = \frac{1}{3}\left(\frac{u^3}{3}\right) + C = \frac{u^3}{9} + C$$

Substituting $u = 3x + 5$, $\int (3x+5)^2 dx = \frac{1}{9}(3x+5)^3 + C$

The correct choice is (D).

6. The average velocity of a particle for $t_1 \le t \le t_2$ is

$$\bar{v} = \frac{x(t_2) - x(t_1)}{t_2 - t_1}$$

where $x(t)$ is the position of the particle at time t.

The position function for this problem is given as $x(t) = \ln t$. Substituting into the expression above gives:

$$\bar{v} = \frac{\ln e - \ln 1}{e - 1} = \frac{1-0}{e-1} = \frac{1}{e-1}$$

The correct choice is (C).

7. Since $e^{xy} = 2$, then $\ln(e^{xy}) = \ln(2)$. Therefore $xy = \ln(2)$ or $y = \frac{\ln 2}{x}$.

$$\frac{dy}{dx} = \frac{-\ln 2}{x^2}$$

At $x = 1$, $\frac{dy}{dx} = -\ln(2)$

The correct choice is (A).

8. Differentiate using the Product Rule:

$f'(x) = 3x\left(\frac{1}{x}\right) + 3\ln x$

Since $3\ln x = \ln(x^3)$, the derivative, $f'(x)$, simplifies to $3 + \ln(x^3)$.

The correct choice is (A).

9. $|x+2| = \begin{cases} x+2, & x \geq -2 \\ -x-2, & x < -2 \end{cases}$

$$\begin{aligned}
\int_{-3}^{3} |x+2|\,dx &= \int_{-3}^{-2} (-x-2)\,dx + \int_{-2}^{3} (x+2)\,dx \\
&= \left(\frac{-x^2}{2} - 2x\right)\Big|_{-3}^{-2} + \left(\frac{x^2}{2} + 2x\right)\Big|_{-2}^{3} \\
&= \left[(-2+4) - \left(-\frac{9}{2}+6\right)\right] + \left[\left(\frac{9}{2}+6\right) - (2-4)\right] \\
&= \left(2 + \frac{9}{2} - 6\right) + \left(\frac{9}{2} + 6 - 2 + 4\right) \\
&= 13
\end{aligned}$$

Alternate Solution:

Sketch the graph of $y = |x+2|$ for $-3 \leq x \leq 3$ and find the areas of the two triangles with bases on the x-axis.

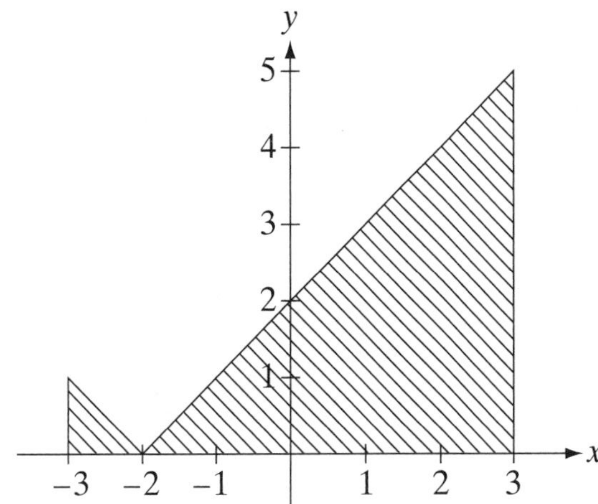

The correct choice is (C).

10.

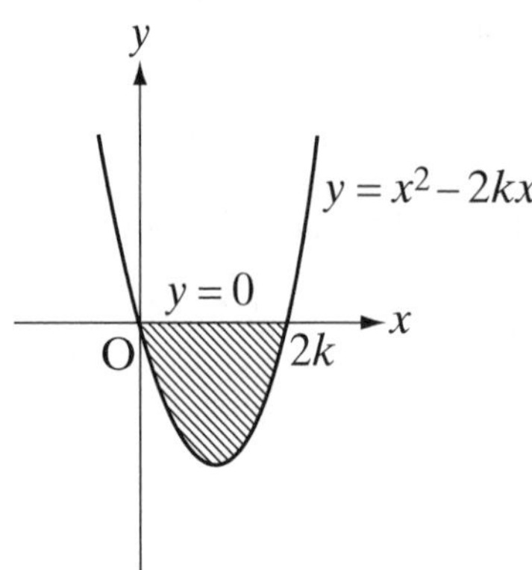

The graph of $y = x^2 - 2kx$ is a parabola whose x-intercepts are $x = 0$ and $x = 2k$.

The x-intercepts are found by setting $y = 0$ and solving for x.

$$x^2 - 2kx = 0$$

$$x(x - 2k) = 0$$

$$x = 0, x = 2k$$

The region R is shown shaded in the figure above. Note: The area of region R is bounded by two functions, $y = 0$ and $y = x^2 - 2kx$. When $0 \le x \le 2k$ the function $y = 0 \ge y = x^2 - 2kx$.

Since the area of the region is given as 36,

$$\int_0^{2k} 0 - (x^2 - 2kx)\,dx = 36$$

$$\int_0^{2k} (-x^2 + 2kx)\,dx = 36$$

$$\left.\frac{-x^3}{3} + \frac{2kx^2}{2}\right|_0^{2k} = 36$$

$$\left.\frac{-x^3}{3} + kx^2\right|_0^{2k} = 36$$

$$\frac{-8k^3}{3} + 4k^3 = 36$$

$$4k^3 = 108$$

$$k^3 = 27$$

$$k = 3$$

The correct choice is (B).

11. $f(x) = e^x - 2x$

$f'(x) = e^x - 2$

$f'(x) = 0$ when $e^x - 2 = 0 \Rightarrow e^x = 2 \Rightarrow x = \ln 2$

Using the First Derivative Test to find possible relative max/min,

	$x < \ln 2$	$x > \ln 2$
f'	$-$	$+$

$x = \ln 2$ is the only relative minimum (where f' switches from – to +).

The minimum value of $f(x)$ is attained when $x = \ln 2$, and the value of the function is

$$f(\ln 2) = e^{\ln 2} - 2(\ln 2) = 2 - 2\ln 2 = 2(1 - \ln 2)$$

The minimum value of the function $f(x) = e^x - 2x$ is $2(1 - \ln 2)$.

The correct choice is (D).

12. Since $f(x)$ is continuous on $[-3,7]$ and differentiable on $(-3,7)$, the Mean Value Theorem guarantees a point c, $-3 < c < 7$, such that $f'(c) = \dfrac{f(7) - f(-3)}{7 - (-3)} = \dfrac{2 - 4}{10} = -\frac{1}{5}$

The correct choice is (D).

13. By the Chain Rule, $h'(x) = f'(g(x))\,g'(x)$.

Thus $h'(2) = f'(g(2))\,g'(2)$.

Then since $f'(x) = 2x + 3$ and $g(2) = 1$, $f'(g(2)) = 2(1) + 3 = 5$ and from the table $g'(2) = 3$. Therefore, $h'(2) = 5(3) = 15$

The correct choice is (D).

14. Write $g(x)$ as $g(x) = [f(x)]^{-1}$, then differentiate using the Power Rule and the Chain Rule. $g'(x) = (-1)[f(x)]^{-2}f'(x)$. Then substitute $x = 2$ and use the values from the table,

$$g'(2) = (-1)[f(2)]^{-2}f'(2) = (-1)(-8)^{-2}(-4) = \frac{4}{64} = \frac{1}{16}$$

<u>Note</u>: You could have used the Quotient Rule to find $g'(x)$.

$$g'(x) = \frac{-f'(x)}{[f(x)]^2};\text{ so } g'(2) = \frac{-f'(2)}{[f(2)]^2} = \frac{4}{64} = \frac{1}{16}$$

The correct choice is (C).

15. $g(15) - g(1) = \int_1^{15} f(t)\,dt$

Since $g(1) = 0$, to approximate $g(15)$ the left Riemann sum adds the function values of f at the left side of each interval multiplied by the width of the interval.

Thus, $g(15) \approx 2(3-1) + 3(6-3) + 4(10-6) + 2(15-10) = 39$

The correct choice is (E).

16. V = Volume of the sphere

S = Surface area of the sphere

$V = \frac{4}{3}\pi r^3$ and $\frac{dV}{dt} = 4\pi r^2 \frac{dr}{dt}$

$S = 4\pi r^2$ and $\frac{dS}{dt} = 8\pi r \frac{dr}{dt}$

Given: $\frac{dV}{dt} = 12$

Find: $\frac{dS}{dt}$ when $V = 36\pi$

Since $\frac{dV}{dt} = 12$, then $12 = 4\pi r^2 \frac{dr}{dt}$ and $\frac{dr}{dt} = \frac{3}{\pi r^2}$

Since $\frac{dr}{dt} = \frac{3}{\pi r^2}$, then $\frac{dS}{dt} = 8\pi r\left(\frac{3}{\pi r^2}\right) = \frac{24}{r}$

Finally, when $V = 36\pi$, $r = 3$ (Since $\frac{4}{3}\pi r^3 = 36\pi \Rightarrow r = 3$)

$$\frac{dS}{dt} = \frac{24}{3} = 8$$

The surface area is increasing at 8 square feet per second.

The correct choice is (A).

17. Substituting for $f(x)$ gives:

$$\begin{aligned}\int_{-2}^{2}[f(x)-g(x)]\,dx &= \int_{-2}^{2}[15-g(x)-g(x)]\,dx\\ &= \int_{-2}^{2}[15-2g(x)]\,dx\\ &= 15\int_{-2}^{2}dx-2\int_{-2}^{2}g(x)\,dx\\ &= 15x\Big|_{-2}^{2}-2\int_{-2}^{2}g(x)\,dx\\ &= 60-2\int_{-2}^{2}g(x)\,dx\end{aligned}$$

Note: Since there is no information about the symmetry of $g(x)$, it cannot be said that $\int_{-2}^{2}g(x)\,dx=2\int_{0}^{2}g(x)\,dx$. Thus choice (D) is not correct.

The correct choice is (E).

18. $\int_{a}^{x}\frac{d}{dt}[f(t)]\,dt=f(t)\Big|_{a}^{x}=f(x)-f(a)$

$\int_{2}^{4}\left[\frac{d}{dt}(3t^2+2t-1)\right]dt=\int_{2}^{4}(6t+2)\,dt=3t^2+2t\Big|_{2}^{4}=(48+8)-(12+4)=40$

Note: Do not confuse $\int_{a}^{x}\frac{d}{dt}[f(t)]\,dt$ with $\frac{d}{dx}\left[\int_{a}^{x}f(t)\,dt\right]$.

$\int_{a}^{x}\frac{d}{dt}[f(t)]\,dt=f(x)-f(a)$ while $\frac{d}{dx}\left[\int_{a}^{x}f(t)\,dt\right]=f(x)$.

The correct choice is (B).

19. $R(t)$, the rate of change of the velocity, is the acceleration. $R(0)=a(0)=-\sqrt{3}$. The velocity $v(0)=1$ (given). Since the velocity is positive, the particle is moving to the right. Since the velocity and accleration have different signs, the speed is decreasing.

The correct choice is (D).

20. Relative extreme values of functions occur where the derivative is zero or undefined (critical points). Since f is differentiable its derivative is defined and must be zero at its relative minimum. I is true.

While f is differentiable, this does not guarantee that the derivative is differentiable, so it is possible that $f''(c)$ does not exist. Hence II is false. Example would be $y = x^{\frac{4}{3}}$.

If the second derivative does exist, it may be zero at the minimum. (For example x^4 has a minimum at the origin, but its second derivative is zero there.) So III is false.

The correct choice is (A).

21. The equation of the circle is $x^2 + y^2 = 9$. The area of an equilateral triangle of side length s is $\frac{\sqrt{3}}{4}s^2$.

The cross section of the solid is an equilateral triangle of side $2y$ and area of $\sqrt{3}\,y^2$ or $\sqrt{3}\,(9 - x^2)$.

$$\begin{aligned}\text{The Volume V } &= \sqrt{3}\int_{-3}^{3}(9 - x^2)\,dx \\ &= \sqrt{3}\left(9x - \frac{x^3}{3}\right)\Bigg|_{-3}^{3} \\ &= \sqrt{3}\,(27 - 9) - \sqrt{3}\,(-27 + 9) \\ &= 36\sqrt{3}\end{aligned}$$

The correct choice is (E).

22. Treating y as an accumulation function, $\int_2^x f(t)\,dt$ gives the accumulated amount from $t = 2$ to $t = x$. Adding to this the amount of 4 already present at $x = 2$ gives $4 + \int_2^x f(t)\,dt$.

Alternately $y(x) - y(2) = \int_2^x f(t)\,dt$ so $y(x) = 4 + \int_2^x f(t)\,dt$.

The correct choice is (B).

23. In general, if $y = \arcsin(u)$, then $y' = \dfrac{du}{\sqrt{1-u^2}}$.

In this problem, $y = \arcsin\left(\frac{3x}{4}\right)$ and let $u = \frac{3x}{4}$

$$y' = \frac{\frac{3}{4}}{\sqrt{1-(\frac{3x}{4})^2}}$$

$$= \frac{\frac{3}{4}}{\sqrt{1-\frac{9x^2}{16}}}$$

$$= \frac{\frac{3}{4}}{\sqrt{\frac{16-9x^2}{16}}}$$

$$= \frac{\frac{3}{4}}{\frac{1}{4}\sqrt{16-9x^2}}$$

$$= \frac{3}{\sqrt{16-9x^2}}$$

The correct choice is (E).

24. Horizontal asymptotes are the graphical manifestations of limits at ∞ and $-\infty$. The following limits will correspond to the values of the horizontal asymptotes.

$\lim\limits_{x\to\infty} \dfrac{ae^x+b}{e^x+1} = \lim\limits_{x\to\infty} \dfrac{a+be^{-x}}{1+e^{-x}} = a$. Hence $a = 3$.

$\lim\limits_{x\to-\infty} \dfrac{ae^x+b}{e^x+1} = b$. Hence $b = -5$.

Therefore, $a + b = -2$.

The correct choice is (C).

25. The slope $\left(\frac{dy}{dx}\right)$ at each point is $\frac{2y}{x}$.

Therefore $\dfrac{dy}{dx} = \dfrac{2y}{x}$ which is a separable differential equation

$$\frac{dy}{dx} = \frac{2y}{x} \Rightarrow \frac{dy}{y} = \frac{2dx}{x} \Rightarrow \int \frac{dy}{y} = 2\int \frac{dx}{x} \Rightarrow \ln y = 2\ln x + C$$

$\ln y = 2\ln x + C \Rightarrow y = e^{2\ln x + C} \Rightarrow y = e^{2\ln x} \cdot e^C$ or $y = Cx^2$, (which graphically is a parabola)

Note: $e^{2\ln x} = e^{\ln(x^2)} = x^2$

The correct choice is (B).

26. The given limit is actually the derivative of tan $2x$.

$$\text{If } f(x) = \tan 2x,\ f'(x) = \lim_{h\to 0}\frac{f(x+h)-f(x)}{h}$$
$$= \lim_{h\to 0}\frac{\tan(2(x+h)) - \tan(2x)}{h}$$

Consequently, the derivative of tan $2x$ is $2\sec^2(2x)$.

The correct choice is (D).

27. The first step is to realize that the area under $y = \sin\left(\frac{x}{2}\right)$ extends from $x = 0$ to $x = 2\pi$. (Note: The period of $\sin\left(\frac{\pi}{2}\right)$ is 4π, so the given graph is one-half of a complete cycle.)

Let $u = \frac{1}{2}x$; $du = \frac{1}{2}dx \Rightarrow dx = 2du$

$$\text{Area} = \int \sin\left(\frac{1}{2}x\right)dx = 2\int \sin(u)\,du = -2\cos\left(\frac{1}{2}x\right)\Big|_0^{2\pi}$$
$$= -2[\cos(\pi) - \cos(0)]$$
$$= -2[(-1) - (1)]$$
$$= -2(-2) = 4$$

The correct choice is (C).

28. By the Fundamental Theorem of Calculus, $f(4) - f(2) = \int_2^4 f'(x)\,dx$. The integral can be evaluated by finding the area of the trapezoid formed by the x-axis, the graph of $f'(x)$ and the lines $x = 2$ and $x = 4$. The area is $\frac{1}{2}(2)(2+3) = 5$. So $f(4) - (-3) = 5$ and $f(4) = 2$.

The integral may also be found by intergrating the equation of the derivative $f'(x)$. Thus,
$\int_2^4\left(\frac{1}{2}x + 1\right)dx = \frac{1}{4}x^2 + x\Big|_2^4 = 4 + 4 - (1 + 2) = 5.$

Then $f(4) - (-3) = 5$ and $f(4) = 2$.

The correct choice is (C).

29. Method I:

The integrals give the area of the region between the x-axis and the graph of the function. Since the region below the x-axis is counted as negative the greatest value will be $\int_0^2 f(x)\,dx$.

Method II:

The choices may be thought of as various values of the function $F(t) = \int_0^t f(x)\,dx$. Since $F'(t) = f(t)$ by the Fundamental Theorem of Calculus, the largest value of F will occur when its derivative, f, changes from positive to negative. From the graph this occurs only at $t = 2$, so $F(2) = \int_0^2 f(x)\,dx$ will be the absolute maximum.

The correct choice is (B).

30. Speed $= |x'(t)| = |4t^3 - 30t^2 + 58t - 36|$. The greatest speed may be found by using a calculator to graph. Otherwise evaluate, $|4t^3 - 30t^2 + 58t - 36|$ for each choice. Of the values given, $t = 4$ gives the greatest speed. (Note: the greatest speed in the interval is at $t \approx 3.690$).

The correct choice is (D).

31. First, observe that since $f'(x) > 0$ for all x, $f(x)$ is increasing. That fact eliminates the graphs of choices (A), (B), and (C). When the first derivative is increasing for $x \le 0$, the second derivative will be positive and function f will be concave upwards. Likewise, when the derivative is decreasing for $x \ge 0$ the function f will be concave downward. This describes graph (E) and eliminates graph (D).

The correct choice is (E).

32. $g(3) = 4(3) + \frac{-6}{3} = 10.$

Then using the Quotient Rule for derivatives:

$$\begin{aligned} g'(x) &= 4 + \frac{xf'(x) - f(x)(1)}{x^2} \\ g'(3) &= 4 + \frac{3(4) - (-6)(1)}{3^2} \\ g'(3) &= 6 \end{aligned}$$

The line through $(3, 10)$ with a slope of 6 is $y - 10 = 6(x - 3)$.

The correct choice is (E).

33. By the Fundamental Theorem of Calculus $f'(t) = e^{t\cos(t)}(\cos(t) - t\sin(t))$. The maximum of f will occur where its derivative changes from positive to negative or at the endpoints of the domain. Graph the derivative in a convenient window, such as x [0,10] by y [-10,10]

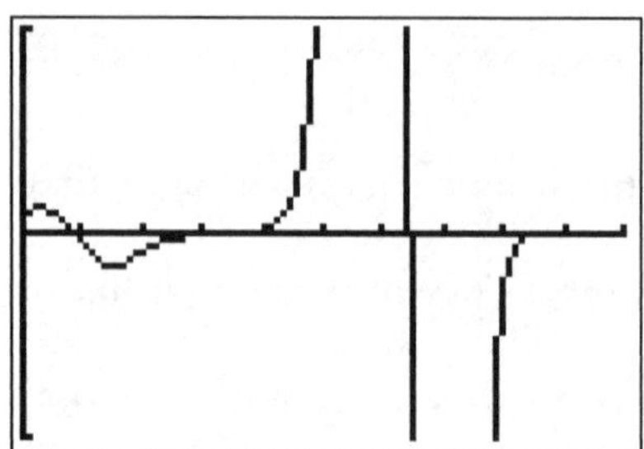

Note that this graph does NOT have a vertical asymptote – this should be obvious from the equation of the derivative.

However it is best to use ZoomFit x [0,10] by y [-922, 854]

The derivative changes from positive to negative to the left of $t = 1$ ($t = 0.860$) and just past 6 ($t = 6.437$). The other zeros are 3.426, and 9.529. Based on the size of the area of the regions between the graph of $f(t)$and the horizontal axis, $t = 6.437$ must be the location of the absolute maximum.

By looking at the answer choices it is *not* necessary to actually calculate the zeros, since only one choice is between 6 and 7.

Alternate Solution:

The integral is of the form $\int e^u du$ where $u = x\cos(x)$ and $du = (\cos(x) - x\sin(x))\,dx$. Therefore $f(t) = e^{x\cdot\cos x}\Big|_0^t = e^{t\cos(t)} - 1$. Graph this and observe that the maximum is between $t = 6$ and $t = 7$.

The correct choice is (C).

34. Distance $= \int_{t_1}^{t_2} v(t)\,dt = \int_{t_1}^{t_2} 2e^{2t}\,dt.$

Find the values of t_1 and t_2 when the velocity is 2 and 4.

$$\begin{cases} 2e^{2t} = 2 \Rightarrow e^{2t} = 1 \text{ or } t = 0 \\ 2e^{2t} = 4 \Rightarrow e^{2t} = 2 \text{ or } t = \frac{1}{2}\ln 2 \end{cases}$$

Thus the distance the particle travels equals $\int_0^{\frac{1}{2}\ln 2} 2e^{2t}\,dt = 1.$

The correct choice is (A).

35. Graph $f'(x)$ in the viewing window $x[0,12]$ by $y[-2,2]$. In the given interval, the derivative changes from positive to negative twice, indicating that $f(x)$ will have two relative maxima (First Derivative Test).

The correct choice is (C).

36. Method I:

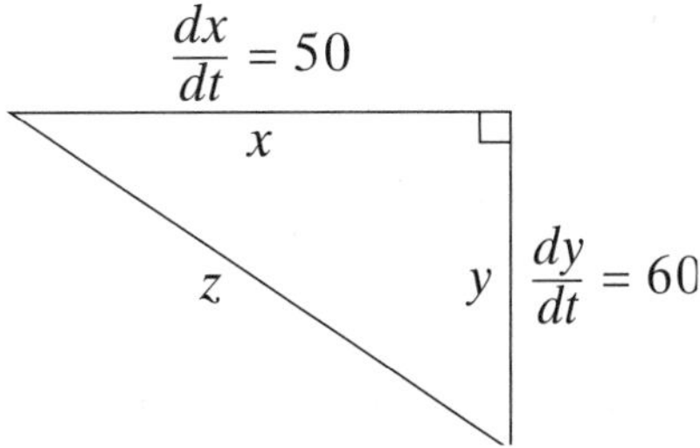

Let x be the distance traveled by the first car and let y be the distance traveled by the second car.

Note that $dx/dt = 50$ and $dy/dt = 60$ and the distance between them is $z = \sqrt{x^2 + y^2}$. (see the above diagram)

Now find dz/dt (the rate at which the distance between them is changing) when $x = 25$ and $y = 30$ (one-half hour later).

Taking the derivative of both sides with respect to t, we get

$$\frac{dz}{dt} = \frac{2x\,\frac{dx}{dt} + 2y\,\frac{dy}{dt}}{2\sqrt{x^2 + y^2}}$$

One-half hour after they start $x = 25$ and $y = 30$. Substitute those values (and the values of dx/dt and dy/dt) to get $dz/dt = 78.102$ or about 78 miles per hour.

Method II:

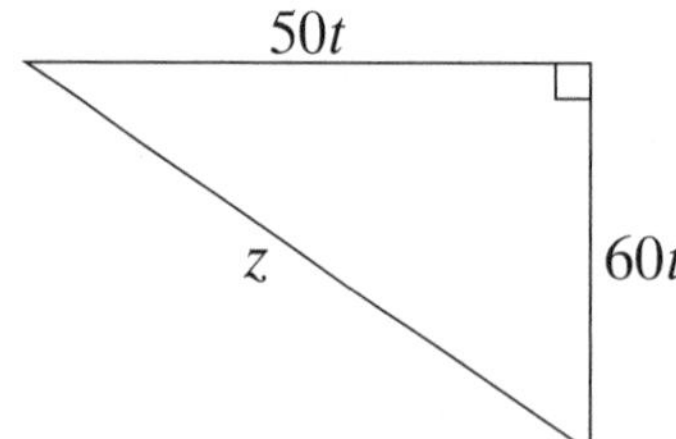

Express the distance traveled as a function of time t. The first car will travel $50t$ miles and the second car will travel $60t$ miles. The distance between them is given by

$$z = \sqrt{(50t)^2 + (60t)^2} \text{ and } \frac{dz}{dt} = \frac{2(50t)(50) + 2(60t)(60)}{2\sqrt{(50t)^2 + (60t)^2}}$$

Substitute $t = \frac{1}{2}$ (one-half hour later) and find that $dx/dt \approx 78$ mph. (Recall that $x = 50t$, $y = 60t$, $dx/dt = 50$; and $dy/dt = 60$).

The correct choice is (D).

37. The area of the square is 4. The area under the parabola $y = -x^2 + 2x$ is given by

$$\int_0^2 (-x^2 + 2x)\, dx = -\frac{x^3}{3} + x^2 \Big|_0^2 = \frac{4}{3}$$

The area above the parabola is $4 - (\frac{4}{3}) = \frac{8}{3}$.

The probability is the ratio of the area above the parabola to the total area: $\frac{\frac{8}{3}}{4} = \frac{2}{3}$.

The correct choice is (C).

38. According to the Fundamental Theorem of Calculus, $\frac{d}{dx}\int_0^x \arcsin(t)\, dt = \arcsin(x)$.

So $\arcsin(0.4) = 0.412$

The correct choice is (C).

39. Method I: By the Fundamental Theorem of Calculus

$g'(x) = (5 + 4x - x^2)(2^{-x}) = -(x - 5)(x + 1)(2^{-x})$ On the interval (3,5), $g' > 0$ so g is increasing (3,5) on, so I is true. On the interval (5,7), $g' < 0$ so g is decreasing on (5,7), so II is false. Use a graphing calculator to find $g(7) = \int_3^7 (5 + 4t - t^2)(2^{-t})\, dt = 0.562$, so III is false.

Method II: Graph the integrand $(5 + 4x - x^2)(2^{-x})$ in a suitable window such as [3,7] by [−1,2] (See figure). This is the derivative of g and from the graph it is clear that the derivative is positive on (3,5) indicating g is increasing, so I is true.

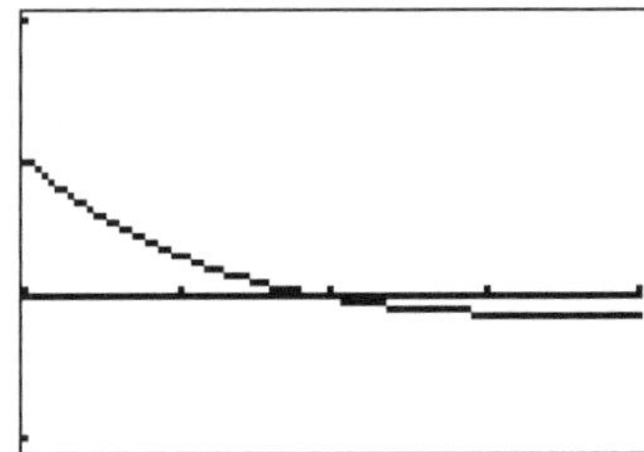

The derivative is negative on (5,7) indicating that g is decreasing, so II is false.

$g(7)$ is the net area between the axis and the graph. Since there is more area above the axis than below, $g(7) > 0$ and III is false.

The correct choice is (A).

40. Volume of the solid is $V = \pi\int_0^1 [(2-x)^2 - (1)^2]\,dx = \pi\int_0^1 (x^2 - 4x + 3)\,dx$

$$= \pi\left(\frac{x^3}{3} - 2x^2 + 3x\right)\Bigg|_0^1 = \frac{4\pi}{3}$$

The correct choice is (D).

41. Find y' and substitute into the differential equations to check them:

$\dfrac{dy}{dx} = \cos x - \sin x$

Equations I and II: $y + \dfrac{dy}{dx} = (\sin x + \cos x) + (\cos x - \sin x) = 2\cos x$ so the equation is not a solution of I and is a solution of II.

Equation III: $\dfrac{dy}{dx} - y = (\cos x - \sin x) - (\sin x + \cos x) = -2\sin x$. This is a solution of III.

The correct choice is (E).

42. Using the disk method, the volume is

$$\pi k^2 \int_0^5 (x-5)^4\,dx = 2500\pi$$

Integrating and solving for k,

$$\pi k^2 \frac{(x-5)^5}{5}\Big|_0^5 = 2500\pi$$

$$\pi k^2 (5^4) = 2500\pi$$

$$k^2 = 4 \text{ or } k = 2$$

The correct choice is (E).

43. Use implicit differentiation to find $\frac{dy}{dx}$:

$$2y\frac{dy}{dx} - 3x^2 - 30x = 0$$

$$2y\frac{dy}{dx} = 3x^2 + 30x$$

$$\frac{dy}{dx} = \frac{3x(x+10)}{2y}$$

The derivative is zero when $x = -10$ and when $x = 0$. These are the values of x for which there are horizontal tangents.

The derivative may also be found by solving for y and then differentiationg:

$$y = \pm\sqrt{x^3 + 15x^2 + 4}$$

$$\frac{dy}{dx} = \frac{3x^2 + 30x}{\pm 2\sqrt{x^3 + 15x^2 + 4}}$$

$$= \frac{3x(x+10)}{\pm 2\sqrt{x^3 + 15x^2 + 4}}$$

Again when $x = -10$ and $x = 0$, the curve will have horizontal tangent lines.

The correct choice is (D).

44. The given limit indicates a horizontal asymptote on the left side of the graph at $y = -3$. Since the function is odd, it is symmetric to the origin and there must be a horizontal asymptote on the right at $y = +3$ and therefore I and III are true. Since the vertical asymptote would indicate a discontinuity and the function is given as continuous there can be no vertical asymptote, II is also true. An example of such a function is $f(x) = \frac{6}{\pi}\tan^{-1}(x)$.

The correct choice is (E).

45. The average value on $[-a,a]$ is given by $\frac{1}{a-(-a)}\int_{-a}^{a} f(x)\,dx$.

Since the interval is symmetric to the origin, only an odd function will have a definite integral whose value is zero. Of the functions given, only $\sin(x)$ is an odd function. This may be confirmed by graphing the function.

Alternately, evaluate each definite integral on a graphing calculator and find which has the value of zero.

The correct choice is (D).

1a. Integrate $v(t)$ to find the position function $x(t)$:

$x(t) = \int (\sin t + e^{-t})\, dt = -\cos t - e^{-t} + C$

Since $x(0) = 0$ from the given information,

$0 = -\cos(0) - e^{-0} + C \Rightarrow 0 = -1 - 1 + C \Rightarrow C = 2$

So, $x(t) = 2 - \cos(t) - e^{-t}$.

1b. The particle is at rest when $v(t) = 0$.

Setting $\sin(t) + e^{-t} = 0$ and using a graphing calculator, $t = \underline{3.183}$.

1c. Using a graphing calculator, the average value of $x(t)$ on $[0,5]$ is:

$$\frac{1}{5-0}\int_0^5 x(t)\, dt = \frac{1}{5}\int_0^5 (2 - \cos t - e^{-t})\, dt = \underline{1.993}$$

Or, exactly,
$$\begin{aligned}\frac{1}{5}\int_0^5 (2 - \cos t - e^{-t} dt) &= \frac{1}{5}(2t - \sin t + e^{-t})\Big|_0^5 \\ &= \frac{1}{5}[(10 - \sin 5 + e^{-5}) - (0 - 0 + 1)] \\ &= \frac{1}{5}(9 - \sin 5 + e^{-5})\end{aligned}$$

1d. From part (b), the particle changes direction (from moving right to moving left) at $t = 3.183$.

Since $x(0) = 0$, $x(3.183) = 2.958$, and $x(5) = 1.710$

the <u>total</u> distance is $|x(3.183) - x(0)| + |x(5) - x(3.183)| = 4.206$

The distance may also be found by evaluating $\int_0^5 |v(t)|\, dt = 4.206$ (using a graphing calculator).

2a. Substitute $L = 10{,}000$ when $d = 40$ into $L = \dfrac{k}{d^2}$ to find k:

$$10{,}000 = \frac{k}{40^2} \Rightarrow = 10^4 \cdot 40^2 \text{ or } 1.6 \times 10^7.$$

2b. From trigonometry, $\dfrac{40}{d} = \cos\theta$, so $d = \dfrac{40}{\cos\theta}$.

Since $L = \dfrac{k}{d^2}$ then $L(\theta) = \dfrac{(1.6)(10^7)}{\left(\frac{40}{\cos\theta}\right)^2} = 10^4\cos^2\theta$

2c. Differentiate $L = 10^4\cos^2\theta$ (from part (b)) with respect to t (time) to find the rate of change of L:

$$\frac{dL}{dt} = 2 \cdot 10^4 \cos\theta(-\sin\theta)\frac{d\theta}{dt}$$

From the given, substitute $\theta = \frac{\pi}{4}$ and $\dfrac{d\theta}{dt} = \dfrac{\pi}{30}$ to find the numerical value of $\dfrac{dL}{dt}$.

$$\begin{aligned}\frac{dL}{dt} &= 2 \cdot 10^4 \cos\left(\tfrac{\pi}{4}\right)\left(-\sin\tfrac{\pi}{4}\right)\left(\tfrac{\pi}{30}\right)\\ &= 2 \cdot 10^4\left(\frac{\sqrt{2}}{2}\right)\left(-\frac{\sqrt{2}}{2}\right)\left(\frac{\pi}{30}\right)\\ &= \frac{-10^4\pi}{30} = -1047.198\end{aligned}$$

Be careful how the answer is phrased: "The strength is decreasing at 1047.198 lumens/sec." or, "The strength is changing at –1047.198 lumens/sec." The negative sign indicates that it is decreasing. Do NOT write "The strength is decreasing at –1047.198 lumens/sec."

REMINDER: Numerical and algebraic answers need not be simplified, and if simplified incorrectly credit will be lost.

3a. Area $= \int_1^5 \left(1 - \frac{1}{x}\right) dx = x - \ln(x)\Big|_1^5$

$$= 5 - \ln(5) - (1 - \ln(1))$$

$$= 4 - \ln(5)$$

3b. The thickness of the washer is dx. The outer radius is $2 - \frac{1}{x}$ and the inner radius is 1.

Volume $= \pi \int_1^5 \left(\left(2 - \frac{1}{x}\right)^2 - (2-1)^2\right) dx$

3c. To find the volume, integrate the given cross-section area by the thickness dx.

$$\int_1^5 (2^x - 2)\, dx = \frac{2^x}{\ln 2} - 2x\Big|_1^5$$

$$= \frac{2^5}{\ln 2} - \frac{2^1}{\ln 2} - 10 - 2$$

$$= \frac{30}{\ln 2} - 12$$

4a. To solve the differential equation, first separate the variables:

$$\frac{dh}{dt} = -k\sqrt{h}$$

$$\frac{dh}{\sqrt{h}} = -k\,dt$$

Integrating both sides:

$$2\sqrt{h} = -kt + C$$

Now, use the initial condition $h(0) = 16$ to find C.

$$2\sqrt{16} = -k(0) + C \Rightarrow C = 8$$

$$2\sqrt{h} = -kt + 8.$$

Now square both sides and solve for h:

$$4 \cdot h(t) = (-kt + 8)^2$$

$$h(t) = \tfrac{1}{4}(8 - kt)^2$$

4b. When $t = 8$, $h = 12.95$;

So, $12.25 = \tfrac{1}{4}(8 - 8k)^2$

Multiply both sides by 4 and take the positive square root (Note that using the negative square root gives a value of $k > 1$):

$$7 = 8 - 8k \Rightarrow k = \tfrac{1}{8} = 0.125$$

4c. The tank will be completely empty when $h(t) = 0$.

$$0 = \tfrac{1}{4}\left(8 - \tfrac{1}{8}t\right)^2$$

Solving for t, $\underline{t = 64 \text{ hours}}$

After 64 hours, the tank will be completely empty.

5a. To find the point(s), first find the slope of chord AB and set it equal to the derivative of f:

Slope of chord AB$= \dfrac{15-(-15)}{-3-3} = -5$

$f'(x) = 4 - 3x^2$

So, $4 - 3x^2 = -5$

$3x^2 = 9$

$x^2 = 3$

$x = \pm\sqrt{3}$

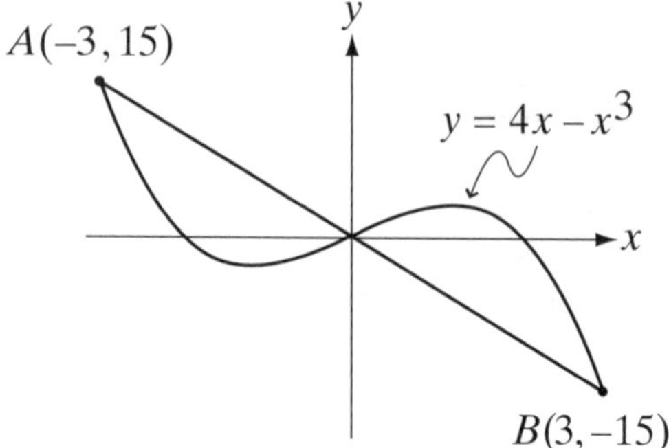

Substitute $x = \sqrt{3}$ and $x = -\sqrt{3}$ into the curve $y = 4x - x^3$ to get $(\sqrt{3}, \sqrt{3})$ and $(-\sqrt{3}, -\sqrt{3})$.

The diagram shows that there are 2 points on the curve where the tangent line is parallel to chord AB.

5b. For $0 < x < 3$, $f(x)$ lies above the chord AB (see diagram above), so the vertical distance is $V(x) = (4x - x^3) - (-5x) = 9x - x^3$.

Note: Chord AB with coordinates $(-3, 15)$ and $(3, -15)$ has an equation of $y = -5x$.

5c. To find the maximum vertical distance, differentiate $V(x)$ and find its zeros.

$V(x) = 9x - x^3$

$V'(x) = 9 - 3x^2$; $V'(x) = 0$ when $9 - 3x^2 = 0 \Rightarrow 3x^2 = 9 \Rightarrow x^2 = 3$ or $x = \sqrt{3}$

(Note: $x = -\sqrt{3}$ is not in the domain $0 < x < 3$)

$x = \sqrt{3}$ is a maximum since $V''(x) = -6x < 0$ at $x = \sqrt{3}$ (Second Derivative Test)

So the maximum vertical distance is $V(\sqrt{3}) = 9\sqrt{3} - (\sqrt{3})^3 = 6\sqrt{3}$

Note: Substitute the value of $x = \sqrt{3}$ into $V(x)$ to find the maximum vertical distance. Do not leave the answer as $x = \sqrt{3}$.

6a. The function will increase when the derivative is non-negative. This occurs when $0 \le x \le 2$ and $3 \le x \le 4$.

6b. The function has a relative minimum value when the derivative changes from negative (on the left) to positive. This occurs at $x = 3$.

6c. The function has a relative maximum value when the derivative changes from positive (on the left) to negative. This occurs at $x = 2$. Note that the derivative does not have to be continuous for this to happen.

6d. Since the derivative is not continuous at $x = 2$ the function will make a sharp turn here. The essential features of the graph are that it is increasing, make a sharp turn at $x = 2$, decrease until $x = 3$, then increase. A possible sketch is:

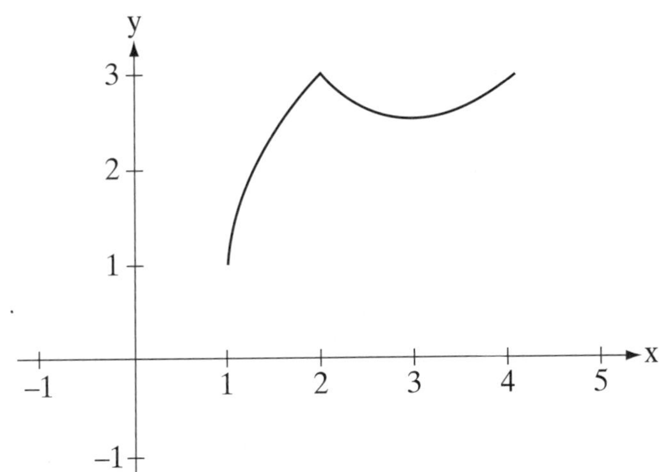

Sample Examination V

1. If $y = (2x^2 + 1)^4$, then $\frac{dy}{dx} = 4(2x^2 + 1)^3(4x) = 16x(2x^2 + 1)^3$.

   The correct choice is (E).

2. Let $u = x^2 + 1$

   $du = 2x\,dx \Rightarrow x\,dx = \frac{1}{2}du$

   $\int x(x^2+1)^{\frac{1}{2}}dx = \frac{1}{2}\int u^{\frac{1}{2}}\,du = \frac{1}{2}\left(\frac{2}{3}u^{\frac{3}{2}}\right) = \frac{1}{3}u^{\frac{3}{2}} = \frac{1}{3}(x^2+1)^{\frac{3}{2}} + C$

   The correct choice is (C).

3. Differentiate the first derivative, using the Chain Rule, to find the second derivative.

$$\begin{aligned}\frac{d^2y}{dx^2} &= 2x\frac{dy}{dx} + 2y\\ &= 2x(2xy) + 2y\\ &= 4x^2y + 2y\end{aligned}$$

   The correct choice is (E).

4. $y = 2x^3 + 24x - 18$

   $y' = 6x^2 + 24 > 0$ for all x

   Therefore, y is increasing for all x.

   The correct choice is (A).

5. $f'(x) = \begin{cases} 2x & \text{for } x \le 1 \\ 2 & \text{for } x > 1 \end{cases}$

   Since $\lim_{x \to 1^-} f'(x) = \lim_{x \to 1^+} f'(x) = 2$, so $f'(1) = 2$.

   The correct choice is (C).

6. $f(x) = 3x^2 - x^3 \Rightarrow f'(x) = 6x - 3x^2$

   To maximize $f'(x)$ take the derivative of $f'(x)$, or $f''(x) = 6 - 6x$.

   Now, $6 - 6x = 0 \Rightarrow x = 1$

	$x < 1$	$x > 1$
f''	+	−

   Therefore, the maximum value of the first derivative of $f(x)$ occurs when $x = 1$, which has a value of 3.

   The correct choice is (D).

7. Statement I is false. Since $\int_0^3 f(x)\,dx$ equals a constant, $\frac{d}{dx}\int_0^3 f(x)\,dx = 0$ by the Fundamental Theorem of Calculus.

   Statement II is false because $\int_3^x f'(x)\,dx = f(x) - f(3)$ (Fundamental Theorem of Calculus).

   Statement III is a true representation of the Fundamental Theorem of Calculus.

   The correct choice is (B).

8. To find $\frac{dy}{dx}$, or y', use implicit differentiation.

$$\begin{aligned} \sin(xy) &= x^2 \\ \cos(xy)(xy' + y) &= 2x \\ xy'\cos(xy) + y\cos(xy) &= 2x \\ xy'\cos(xy) &= 2x - y\cos(xy) \\ y' &= \frac{2x - y\cos(xy)}{x\cos(xy)} \end{aligned}$$

   Now, since $\frac{1}{\cos(xy)} = \sec(xy)$, $y' = \frac{2x\sec(xy) - y}{x}$

   The correct choice is (E).

9. The graph of a function changes from concave down to concave up where its second derivative changes from negative to positive. This will occur where the first derivative changes from decreasing to increasing. The first derivative changes from decreasing to increasing at its local minimum points. Thus, the graph of g changes from concave down to concave up where g' has local minimums.

   The correct choice is (D).

10. The graph of $y = 2x^6 + 9x^5 + 10x^4 - x + 2$ has a point of inflection where the second derivative changes sign.

$$\begin{aligned} y' &= 12x^5 + 45x^4 + 40x^3 - 1 \\ y'' &= 60x^4 + 180x^3 + 120x^2 \\ &= 60x^2(x^2 + 3x + 2) \\ &= 60x^2(x+2)(x+1) \Rightarrow x = 0, -1, -2 \text{ are candidates for points of inflection} \end{aligned}$$

Test whether the second derivative changes sign at these points.

	$x < -2$	$-2 < x < -1$	$-1 < x < 0$	$x > 0$
f''	+	−	+	+

Therefore, only $x = -2$ and $x = -1$ are points of inflection.

The correct choice is (C).

11. Since the function is increasing the left Riemann sum rectangles will all lie below the graph of the function and therefore their sum is less than the area between the graph and the x-axis, $\int_1^4 f(x)\,dx$. All of the other choices have part or all of their areas above or partly above the graph and will be larger than the left Riemann sum.

The number of subintervals in this situation does not matter. To compare the relative sizes draw one rectangle or trapezoid on the interval [1,4]. The relative sizes of these rectangles will be true for any number of rectangles.

The correct choice is (B).

12. Note the cyclic pattern of derivatives of $y = \sin(2x)$:

$$\begin{aligned} y &= \sin(2x) \\ y' &= 2\cos(2x) = 2^1\cos(2x) \\ y'' &= -4\sin(2x) = -2^2\sin(2x) \\ y''' &= -8\cos(2x) = -2^3\cos(2x) \\ y^{(4)} &= 16\sin(2x) = 2^4\sin(2x) \\ y^{(5)} &= 32\cos(2x) = 2^5\cos(2x) \end{aligned}$$

Each time a derivative is taken, the Chain Rule introduces another factor of 2.

Also note that every "even" derivative of $y = \sin(2x)$ contains $\sin(2x)$ as a factor. So the 20th derivative will contain $\sin(2x)$ as a factor, eliminating choices (C), (D), and (E). The 20th derivative is "similar" to the 4th derivative.

The correct choice is (B).

13. Begin by finding where $f'(x) = 3$:

$$\begin{aligned} f'(x) &= 7 - 2x \\ 3 &= 7 - 2x \\ -4 &= -2x \\ 2 &= x \end{aligned}$$

Find the point of tangency: $f(2) = 7(2) - 2^2 = 10$. The equation of the line through $(2,10)$ with a slope of 3 is

$$\begin{aligned} y - 10 &= 3(x - 2) \\ y - 10 &= 3x - 6 \\ y &= 3x + 4 \end{aligned}$$

The correct choice is (B).

14. Statement I needs not be true. Let $f(x) = x^3$ on $[-1,2]$. $f'(0) = 0$ but $f(-1) \neq f(2)$.

Statement II also needs not be true. Again, consider $f(x) = x^3$ on $[-1,2]$. $f'(0) = 0$ but $x = 0$ is not a relative extremum (maximum or minimum).

Statement III also needs not be true. Consider $f(x) = x^3 + 2x^2$. Although $f'(0) = 0$, $f''(0) \neq 0$.

The correct choice is (A).

15. Multiply the function values at the right side of each interval by the width of the interval. Add these four products to determine the value of the right Riemann sum.

Thus, $0(3-1) + 3(7-3) + 3(8-7) + (-4)(10-8) = 7$.

The correct choice is (C).

16. The quickest way to do this problem is to simplify $f(x) = x\sqrt[3]{x} = x\left(x^{\frac{1}{3}}\right) = x^{\frac{4}{3}}$.

Therefore, $f'(x) = \frac{4}{3}x^{\frac{1}{3}}$.

The correct choice is (C).

17. Let $u = x + 2 \Rightarrow du = dx$

$$\int_k^6 \frac{dx}{x+2} = \int \frac{du}{u} = \ln|u| = \ln|x+2|\Big|_k^6 = \ln 8 - \ln(k+2) = \ln\left(\frac{8}{k+2}\right).$$

Since $\int_k^6 \frac{dx}{x+2} = \ln k$, $\ln\left(\frac{8}{k+2}\right) = \ln k$, or $\frac{8}{k+2} = k$.

Solving for k, $k^2 + 2k = 8 \Rightarrow k = 2, -4$. Since $k > 0$, the required value of k is 2.

The correct choice is (B).

18. Let $u = \ln x \Rightarrow du = \frac{dx}{x}$

$$\int \frac{dx}{x \ln x} = \int \frac{du}{u} = \ln|u| = \ln|\ln x|\Big|_e^{e^2} = \ln(\ln e^2) - \ln(\ln e) = \ln 2 - \ln 1 = \ln 2$$

<u>Note</u>: $\ln e^2 = 2$, $\ln e = 1$, and $\ln 1 = 0$

The correct choice is (A).

19. $g(x) = f(3x)$, then using the Chain Rule, $g'(x) = f'(3x)(3)$.

Therefore, $g'(0.1) = f'(0.3)(3) = (1.096)(3) = 3.288$

The correct choice is (E).

20. The derivative of $\frac{8x+k}{x^2}$ is $\frac{8x^2 - 2x(8x+k)}{x^4}$ or $\frac{-8x-2k}{x^3}$.

Since $\frac{8x+k}{x^2}$ has a relative maximum at $x = 4$, $\frac{-8x-2k}{x^3} = 0$, or $k = -4x$.

At $x = 4$, $k = -16$.

The correct choice is (B).

21. First recognize the limit as the derivative of the function $f(x) = 2x^5 - 5x^3$.

By definition, $f'(x) = \lim_{h \to 0} \frac{f(x+h) - f(x)}{h} = \lim_{h \to 0} \frac{[2(x+h)^5 - 5(x+h)^3] - [2x^5 - 5x^3]}{h}$

$$= \lim_{h \to 0} \frac{2(x+h)^5 - 5(x+h)^3 - 2x^5 + 5x^3}{h}$$

Therefore the answer is $f'(x) = 10x^4 - 15x^2$.

The correct choice is (D).

22. Using one of the properties of the definite integral, $\int_2^4 f(x)\,dx + \int_4^8 f(x)\,dx = \int_2^8 f(x)\,dx$.

Or, $6 + \int_4^8 f(x)\,dx = -10 \Rightarrow \int_4^8 f(x)\,dx = -16$.

Therefore $\int_8^4 f(x)\,dx = -\int_4^8 f(x)\,dx = -(-16)$ or 16.

Note: $\int_a^b f(x)\,dx = -\int_b^a f(x)\,dx$.

The correct choice is (E).

23. By the Fundamental Theorem of Calculus, $g' = f$ and $g'' = f'$. Points of inflection of g will occur where its second derivative changes sign. From the table, $f'(x)$ changes sign at $x = 2$ and $x = 6$. Thus g has points of inflection at those two values of x.

The correct choice is (D).

24. The inverse of $y = x^{-\frac{1}{3}}$ is $x = y^{-\frac{1}{3}} \Rightarrow y = x^{-3}$.

Therefore, the derivative of $y = x^{-3}$ is $y' = -3x^{-4}$.

The correct choice is (E).

25. Statements I and II must be true by the Extreme Value Theorem.

Statement III is not necessarily true. Consider $f(x) = x^2$ on $[1,4]$. $f'(c) \neq 0$ for $1 < c < 4$.

The correct choice is (C).

26. Picture the graph on the slope field. Look for places where the segments are not parallel to the graph and eliminate those choices. Note the changes in the slope of the segments in the third and fourth quadrants.

The correct choice is (C).

27. Let x = the length of each edge of the cube

S = the surface area of the cube ($S = 6x^2$)

V = the volume of the cube ($V = x^3$)

Given: $\frac{dV}{dt} = 20$ when $x = 10$ Find: $\frac{dS}{dt}$ when $x = 10$

Since $S = 6x^2$ and $V = x^3 \Rightarrow x = V^{\frac{1}{3}}$ and $S = 6V^{\frac{2}{3}}$.

$\frac{dS}{dt} = 4V^{-\frac{1}{3}}\frac{dV}{dt} \Rightarrow \frac{dS}{dt} = \frac{4}{x}\left(\frac{dV}{dt}\right)$

Therefore, $\frac{dS}{dt} = \left(\frac{4}{10}\right)(20) = 8$

The correct choice is (E).

28. Solve the differential equation by separating the variables and integrating both sides of the equation:

$y\,dx = x\,dx$

$y^2 = x^2 + C$ Substitute the initial condition $y(3) = 4$ to find C.

$4^2 = 3^2 + C \Rightarrow C = 7$

Therefore, $y^2 = x^2 + 7$ or $x^2 - y^2 = -7$

The correct choice is (A).

29. An analysis of the derivative, $f'(x)$, may be summarized by the number line

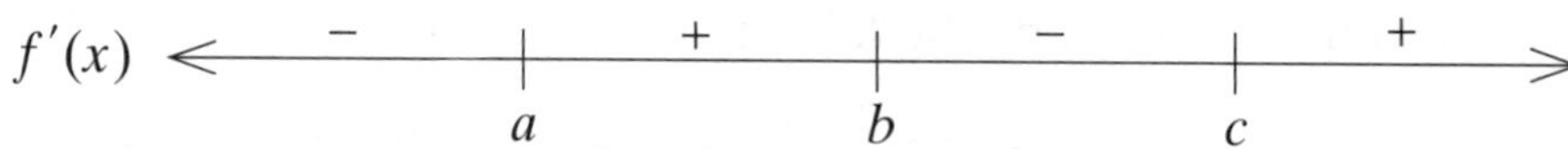

From the First Derivative Test: moving from left to right, f decreases to a minimum at $x = a$, increases to a maximum at $x = b$, decreases to a minimum at $x = c$, and then increases thereafter. This is graph (B).

The correct choice is (B).

30. First find the intersection of the graphs by solving $3\cos x = x$. Using a graphic solution gives $x = 1.170$. Then $A = \int_0^{1.170} (3\cos x - x)\,dx = 2.078$

The correct choice is (C).

31. I is the integral which will give the area using vertical rectangles. II and III are identical except for the "dummy" variable of integration; they both represent the area found using horizontal rectangles.

    The correct choice is (E).

32. The horizontal asymptote is found by taking $\lim\limits_{x\to+\infty} \dfrac{2-e^{\frac{1}{x}}}{2+e^{\frac{1}{x}}}$ and $\lim\limits_{x\to-\infty} \dfrac{2-e^{\frac{1}{x}}}{2+e^{\frac{1}{x}}}$.

    Thus, $\lim\limits_{x\to+\infty} \dfrac{2-e^{\frac{1}{x}}}{2+e^{\frac{1}{x}}} = \dfrac{2-e^0}{2+e^0} = \dfrac{2-1}{2+1} = \dfrac{1}{3}$

    $\lim\limits_{x\to-\infty} \dfrac{2-e^{\frac{1}{x}}}{2+e^{\frac{1}{x}}} = \dfrac{2-e^0}{2+e^0} = \dfrac{2-1}{2+1} = \dfrac{1}{3}$

    The correct choice is (C).

33. Graph $f'(x)$ on $[2,10]$, and note that:

    - $f'(x)$ changes sign twice, so that the function decreases, then increases, and then decreases again. Therefore, I is false.
    - $f'(x)$ changes from negative to positive only once, so II is true.
    - $f'(x)$ has three turning points, so $f''(x)$ will have three zeros and $f(x)$ will have three points of inflection. Therefore, III is true.

    The correct choice is (D).

34. Graph the function in a suitable window such as $x[-2,5]$ by $y[-5,15]$; The graph is tangent to the x-axis at $x = 2$.

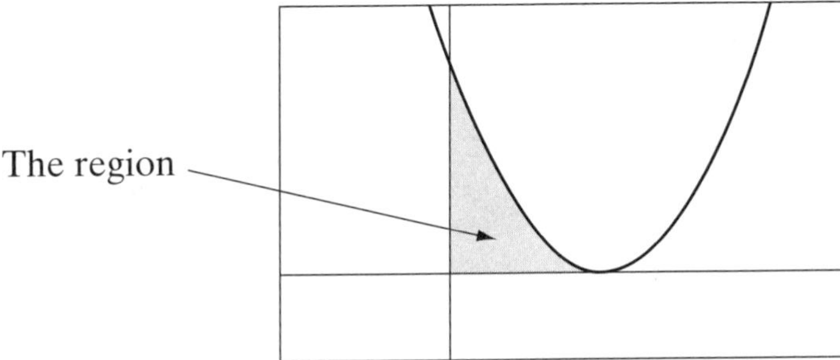

The side of each square cross section lies in the xy-plane and its length is the y-coordinate of the point on the graph; its area is y^2.

The thickness of each piece is in the x-direction and is represented by dx.

The total volume may be found by using the definite integral to sum the individual pieces.

$\int_0^2 (3(x-2)^2)^2 dx = 57.6$ using a graphing calculator.

The correct choice is (E).

35. The rate of change of $\sin^2 x$ is given by

$$\frac{d}{dt}(\sin^2 x) = 2 \sin x \cos x \frac{dx}{dt}$$

When $x = \frac{\pi}{4}$,

$$\begin{aligned} \frac{d}{dt}(\sin^2 x)\Big|_{x=\pi/4} &= 2 \sin\frac{\pi}{4} \cos\frac{\pi}{4} \frac{dx}{dt} \\ k\frac{dx}{dt} &= 2 \cdot \frac{\sqrt{2}}{2} \cdot \frac{\sqrt{2}}{2} \frac{dx}{dt} \\ k\frac{dx}{dt} &= \frac{dx}{dt} \\ k &= 1 \end{aligned}$$

The correct choice is (C).

36. By the Fundamental Theorem of Calculus: $\int_2^5 \frac{dy}{dx} dx = y(5) - y(2) = 50 - 4 = 46$

The correct choice is (A).

37. The average value of $f(x)$ over the given interval is $\frac{1}{\pi - 1}\int_1^\pi e^{-x} \sin x \, dx = 0.129$.

The correct choice is (A).

38. The answer can be determined by looking at the graph of the position, the velocity, or the acceleration.

Method 1: Graph the **position** equation in a suitable window such as $x[0,5]$ by $y[-0.5,0.5]$. The acceleration changes from negative to positive where the position equation changes from concave down to concave up. This happens three times as indicated by the arrows.

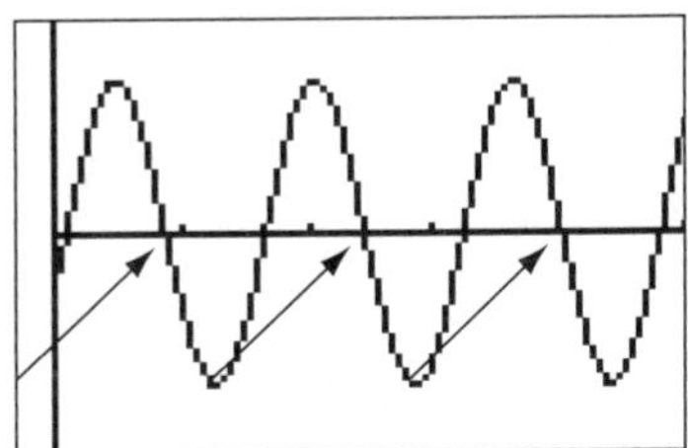

Method 2: The **velocity** is $v(t) = y'(t) = \frac{4}{3}\cos(4t) + \frac{1}{2}\sin(4t)$. Graph the velocity in a suitable window such as $x[0,5]$ by $y[-1.5,1.5]$. Since the acceleration is the derivative of the velocity, the acceleration will change from negative to positive at the values where the velocity changes from decreasing to increasing. This happens three times as indicated by the arrows in the sketch below.

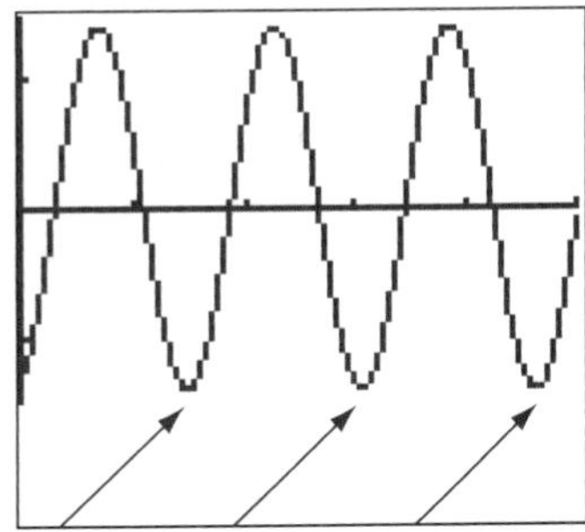

Method 3: The **acceleration** is $a(t) = v'(t) = y''(t) = -\frac{16}{3}\sin(4t) + 2\cos(4t)$. Graph in a suitable window such as $x[0,5]$ by $y[-6,6]$. Count the times the acceleration changes from negative to positive. This happens three times as indicated by the arrows.

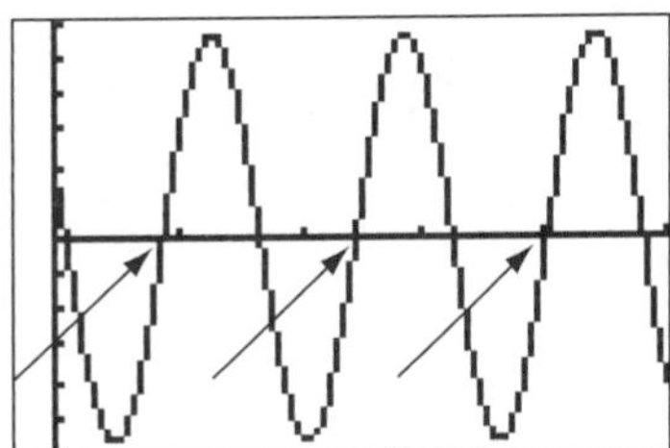

The correct choice is (A).

39. Let $p(t) =$ price of the car at time t

$p'(t) =$ rate of change of the price of the car

It is given that $p'(t) = 120 + 180\sqrt{t}$.

Thus $p(t) = \int p'(t)\,dt = \int (120 + 180\sqrt{t}\,)\,dt = 120t + 120t^{\frac{3}{2}} + C$

Since $p(0) = 14{,}500$, so $C = 14{,}500$.

Therefore $p(t) = 120t + 120t^{\frac{3}{2}} + 14{,}500 \Rightarrow p(5) = 120(5) + 120(5)^{1.5} + 14{,}500 \approx 16{,}440$.

This may also be approached as an accumulation function.

$p(5) = 14{,}500 + \int_0^5 (120 + 180\sqrt{t}\,)\,dt \approx 16{,}440$

The correct choice is (C).

40. <u>Method I</u>: Solve the differential equation by separating the variables:

$$\begin{aligned} \frac{dy}{y} &= kdt \\ \ln|y| &= kt + C_1 \\ |y| &= e^{kt + C_1} \\ y &= Ce^{kt} \text{ where } C = e^{C_1} > 0 \end{aligned}$$

When $t = 0$ then $y = y(0)$

$$\begin{aligned} y(0) &= Ce^0 \\ y(0) &= C \end{aligned}$$

Substitute the given condition that $y(20) = \frac{1}{2}y(0)$, and solve for k:

$$\begin{aligned} \frac{1}{2}y(0) &= y(0)\,e^{20k} \\ \frac{1}{2} &= e^{20k} \\ \ln\left(\frac{1}{2}\right) &= 20k \\ \frac{\ln\left(\frac{1}{2}\right)}{20} &= k \\ -0.035 &= k \end{aligned}$$

Method II: The equation $y(t) = y(0)\, e^{-\frac{\ln 2}{T}t}$ where T is the half-life (here 20 days) models this situation. Then $k = -\dfrac{\ln 2}{20} = -0.035$.

The correct choice is (D).

41. Using the Product Rule, the slope is $f'(x) = x^2\left(\frac{1}{x}\right) + 2x \ln x = x + 2x \ln x$. To find where the slope is 2 use your calculator to solve $x + 2x \ln x = 2$. Thus $x = 1.305$.

The correct choice is (A).

42. The fifth week extends from $t = 4$ to $t = 5$. The average rate of change is given by

$$\frac{m(5) - m(4)}{5 - 4} = \frac{0.92758 - 0.27299}{1} = 0.655$$

The values are found by evaluating the function with a graphing calculator.

The correct choice is (D).

43. I is true since $f'(2) > 0$. Somewhere in the interval $-1 \le x \le 1$ the derivative changes from negative to positive indicating a relative minimum, so II is true. $f(0)$ is greater than $f(-1)$ and $f(1)$, so there is a relative maximum; thus III is true.

The correct choice is (E).

44. The velocity $v(t) = x'(t) = 3t^2 + 22 + 6\pi \sin(\pi t)$. From the graph of $v(t)$ it should be clear that the derivative $x'(t)$ or $v(t)$ is never negative. Alternately, $6\pi \sin(\pi t)$ is < 22, and $3t^2$ is always ≥ 0. Therefore, $v(t) = 3t^2 + 22 + 6\pi \sin(\pi t)$ is always positive.

The correct choice is (E).

45. $\int_0^{-3} f(t)\, dt = -\int_{-3}^{0} f(t)\, dt = -6$. This is the opposite of the area of the triangle. Therefore:

$$\begin{aligned} g(-3) &= g(0) + \int_0^{-3} f(t)\, dt \\ 2 &= g(0) + (-6) \\ 8 &= g(0) \end{aligned}$$

The correct choice is (E).

1a. Graph, $A(t)$ and $B(t)$.

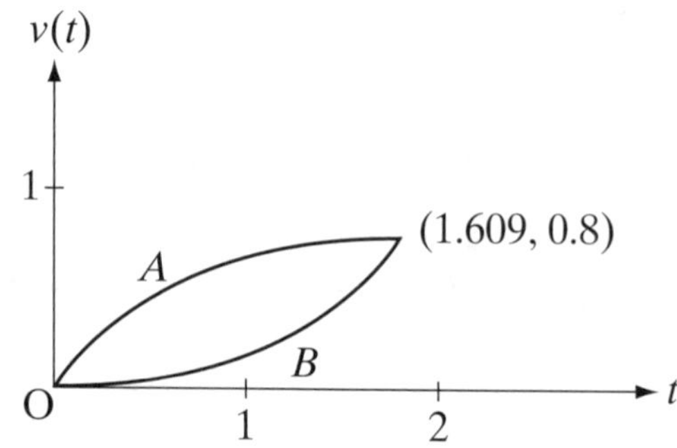

Because of the way the question is worded ("They travel until they have the same velocities."), the graph should end when the graphs intersect.

1b. Person A.

The graphs are velocities, therefore, the area under them represents the distance traveled. Since the area under the graph of A is larger, person A has traveled farther. Alternately, each distance could be calculated (A's distance is $\int_0^{1.609} (1 - e^{-t})\, dt$ see (c) and (d) below) and the answers compared).

1c. $\int_0^x ((1 - e^{-t}) - 0.2(e^t - 1))\, dt$

If x represents the time traveled, then $\int_0^x ((1 - e^{-t}) - 0.2(e^t - 1))\, dt$ represents the distance between the walkers as a function of time.

1d. At $t = 1.609$ minutes

The distance is greatest when the velocities are equal; after this time, B will be walking faster and gaining on A. This occurs when the graphs meet at $t = 1.609$ minutes. At this point, the area under the graph of B begins increasing faster than the area under A.

Alternately, the greatest distance will occur when the derivative of (c), $(1 - e^{-t}) - 0.2(e^t - 1)$ is equal to zero. Solving this equation gives $t = 0$, a minimum (when they start), and $t = 1.609$ minutes, a maximum (when B begins to travel faster than A).

Graph $f'(x)$ in the window $x[0,2]$ and $y[-2,2]$.

2a. The extrema occur where the derivative is zero. These may be found by graphing $f'(x)$, but be careful — there are <u>two</u> zeros of $f'(x)$ near $x = 1.5$ Recommendation: graph $f'(x)$ in the window x:[0,2] by y:[−0.1,0.1].

One zero is where $\cos x = 0$ at $x = \frac{\pi}{2} = 1.571$. The second zero is one of the solutions of $x^3 - 2x = x(x^2 - 2) = 0$ or $x = 0$ and $x = \sqrt{2} - 1.414$. The latter, $x = \sqrt{2}$, is a relative minimum. The relative maximum occurs at $x = \frac{\pi}{2}$ since this is the point where the derivative changes from positive to negative (First Derivative Test).

<u>Note</u>: Do not consider $x = -\sqrt{2}$, as $-\sqrt{2}$ is not in the given domain $0 \le x \le 2$.

2b. The points of inflection occur where the second derivative, $f''(x)$, is zero; this is where the first derivative has a horizontal tangent. Check the derivative on the calculator to find these values, or use the calculator's equation solving ability to solve
$f''(x) = -(x^3 - 2x)(\sin x) + (3x^2 - 2)(\cos x) = 0$. The values are $x = 0.642$ and $x = 1.496$.

2c. There are two possible choices, $x = 0$ or $x = \frac{\pi}{2}$ (from part (a)).

Since most of the graph of the derivative lies below the x-axis,

$f\left(\frac{\pi}{2}\right) - f(0) = \int_0^{\frac{\pi}{2}} (x^3 - 2x)(\cos x)\, dx$ is negative.

Thus, $f\left(\frac{\pi}{2}\right) < f(0)$ and the function's absolute maximum value occurs at $x = 0$.

REMINDER: Numerical and algebraic answers need not be simplified, and if simplified incorrectly credit will be lost.

3a. $g(3) = -3 + \int_2^3 f(t)\,dt = -3 + \frac{1}{2} = -2.5$

$g'(3) = -1 + f(3) = -1 + 1 = 0$

3b. $g'(x) = -1 + f(x) = 0$

$f(x) = 1$

From the graph, you can see that $f(x)$ will equal 1 when $x = 1, 3, 4.5$

3c. Since $g'(x) = -1 + f(x)$, then $g''(x) = f'(x)$.

3d.

x	$g'(x)$	$g''(x)$	Conclusion
1	0	-1	Relative maximum by the Second Derivate Test
3	0	1	Relative minimum by the Second Derivative Test
4.5	0	-2	Relative maximum by the Second Derivative Test

4a. The midpoints of the 5 intervals are $t = 0.1, 0.3, 0.5, 0.7,$ and 0.9. The midpoint sum is

$$(0.2)(90 + 80 + 80 + 40 + 10) = 60$$

4b. The antiderivative of acceleration is the velocity. Thus $f(1)$ is the velocity of the train at the end of the first hour. The units are miles/hour.

4c. $$v(t) = \int a(t)\,dt = \int 90\,dt$$

$$v(t) = 90t + C$$

$$v(0) = 0 = 90(0) + C$$

$$0 = C$$

$$s(t) = \int_{0.1}^{0.2} v(t)\,dt = \int_{0.1}^{0.2} 90t\,dt$$

$$s(t) = 1.35 \text{ miles}$$

5a.

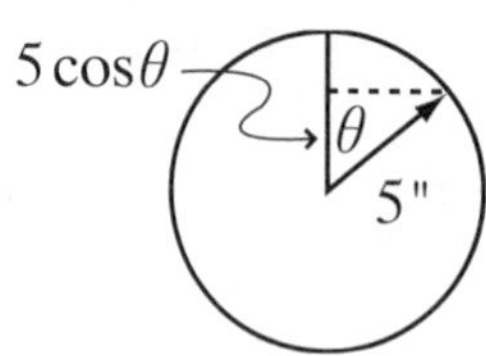

The line indicated in the figure has a length of $5 \cos \theta$. Add this to 35 inches that the center is above the floor to get the expression $y = 35 + 5 \cos \theta$.

5b. Since the hand travels through 2π radians each 60 minutes, the change in θ is $\frac{d\theta}{dt} = 2\pi \cdot \frac{1}{60} = \frac{\pi}{30}$ radians per minute. By integration, $\theta = \frac{\pi}{30}t$. (Since θ and t both start at zero, the constant of integration is zero.)

5c. Differentiate your answer to (a) with respect to t and substitute for $\frac{d\theta}{dt}$.

$$\frac{dy}{dt} = -5 \sin \theta \cdot \frac{d\theta}{dt} = -5 \sin \theta \cdot \frac{\pi}{30} = -\frac{\pi}{6} \sin \theta$$

5d. The answer to (c) gives the rate of change in y. To find when the rate of change is largest (when the increase is most rapid), find its derivative $y'' = -\frac{\pi}{6} \cos \theta \frac{d\theta}{dt}$. Since $\frac{d\theta}{dt}$ is constant, y'' will be zero when $\cos \theta = 0$. This occurs when $\theta = \frac{\pi}{2}$ and $\frac{3\pi}{2}$ or 15 and 45 minutes after the hour. The distance is increasing when the hand is moving upwards at 45 minutes after the hour.

6a. Separate the variables and solve the differential equation:

$$\frac{dT}{dt} = 14 - 0.01T$$

$$\frac{dT}{14 - 0.01T} = dt$$

$$-100 \ln(14 - 0.01T) = t + k$$

$$\ln(14 - 0.01T) = -0.01t + k$$

$$14 - 0.01T = k \cdot e^{-0.01t}$$

$$T(t) = -100(k \cdot e^{-0.01t} - 14)$$

$$T(t) = 1400 - k \cdot e^{-0.01t}$$

6b. Substitute $t = 0$ and $T = 0$ to find k:

$$0 = 1400 - k \cdot e^{0}$$

$$0 = 1400 - k(1)$$

$$k = 1400$$

6c. The temperature is given by $T(t) = 1400 - 1400e^{-0.01t}$.

As $t \to \infty$, that is, if the burner is left on a long time, $\lim_{t \to \infty} T(t) = 1400$.

The hottest the burner will ever get is about 1400°F. Since this is less than 1400°F, the burner is safe.

Note: $\lim_{t \to \infty}(1400 - 1400e^{-0.01t}) = \lim_{t \to \infty} 1400(1 - e^{-0.01t}) = 1400$

Sample Examination VI

1. The segments in the slope field will be horizontal when the numerator $x^2y + y^2x = xy(y + x) = 0$. This occurs when $x = 0$, or $y = 0$, or $y = -x$.

   The correct choice is (E).

2. This is the integration of a basic derivative. Since the derivative of $\cot x$ is $-\csc^2 x$, so $\int \csc^2 x \, dx = -\cot x + C$.

   The correct choice is (E).

3. $f(x) = (x - 1)^2 \cos x$

   Using the Product Rule,

   $$f'(x) = (x - 1)^2 (-\sin x) + (\cos x)(2(x - 1))$$
   $$f'(0) = (0 - 1)^2 (-\sin 0) + (\cos 0)(2(0 - 1))$$
   $$= 0 + (1)(-2)$$
   $$= -2$$

   The correct choice is (A).

4. If $y = \frac{3x + 4}{4x - 3}$, then by the Quotient Rule, $y' \frac{3(4x - 3) - 4(3x + 4)}{(4x - 3)^2}$, or $y' = \frac{-25}{(4x - 3)^2}$. When $x = 1$, $y' = -25$.

   The equation of the tangent line is $y - 7 = -25(x - 1)$ or $y + 25x = 32$.

   The correct choice is (A).

5. The acceleration of a particle $a(t) = v'(t)$ where $v(t) = x'(t)$.

   $$x(t) = \frac{1}{2} \sin t + \cos(2t)$$
   $$v(t) = x'(t) = \frac{1}{2} \cos t - 2 \sin(2t)$$
   $$a(t) = v'(t) = -\frac{1}{2} \sin t - 4 \cos(2t)$$

   Therefore $a(\frac{\pi}{2}) = -\frac{1}{2} \sin(\frac{\pi}{2}) - 4\cos(\pi) = -\frac{1}{2} - 4(-1) = \frac{7}{2}$

   The correct choice is (E).

6. As $x \to +\infty$, both $\frac{1}{2x}$ and $\frac{1}{6x}$ approach 0.

   So, $\lim_{x \to +\infty} \frac{x - \frac{1}{2x}}{2x + \frac{1}{6x}} = \lim_{x \to +\infty} \frac{x}{2x} = \lim_{x \to +\infty} \frac{1}{2} = \frac{1}{2}$

   The correct choice is (D).

7. If the line $y = 4x + 3$ is tangent to the curve $y = x^2 + c$, then the slope of the line $y = 4x + 3$, which is 4, has to be equal to the slope of the tangent line to the curve $y = x^2 + c$ (which is $2x$, the derivative of $x^2 + c$). Since $2x = 4$ or $x = 2 \Rightarrow y = 11$ since $y = 4x + 3$.

   Therefore, to find c, substitute $x = 2$ and $y = 11$ into $y = x^2 + c \Rightarrow c = 7$.

   The correct choice is (C).

8. Let $u = x^2 + 16$, so $du = 2x\,dx \Rightarrow x\,dx = \frac{1}{2}du$

   $$\int_0^3 \frac{x}{\sqrt{x^2 + 16}} dx = \frac{1}{2}\int \frac{du}{\sqrt{u}} = \frac{1}{2}\int u^{-\frac{1}{2}} du = \frac{1}{2}(2u^{\frac{1}{2}}) = u^{\frac{1}{2}} = (x^2 + 16)^{\frac{1}{2}}$$

   Therefore, $(x^2 + 16)^{\frac{1}{2}} = \sqrt{x^2 + 16}\Big|_0^3 = \sqrt{25} - \sqrt{16} = 5 - 4 = 1$.

   The correct choice is (A).

9. $y = \ln(3x + 5)$

   $\frac{dy}{dx} = \frac{3}{3x + 5}$ $\left(\text{The derivative of } \ln u \text{ is } \frac{du}{u}\right)$

   $\frac{d^2y}{dx^2} = \frac{-9}{(3x + 5)^2}$

   The correct choice is (D).

10. The graphs intersect when $x^3 = x \Rightarrow x^3 - x = 0 \Rightarrow x(x^2 - 1) = 0 \Rightarrow x = 0, -1, 1$.

    Since $x^3 \geq x$ on $[-1, 0]$ and $x \geq x^3$ on $[0, 1]$, the total area is

    $$\int_{-1}^{0} (x^3 - x)\,dx + \int_0^1 (x - x^3)\,dx = \frac{1}{4} + \frac{1}{4} = \frac{1}{2}.$$

    The correct choice is (B).

11. By implicit differentiation, $2yy' - [(2x)(y') + 2y] = 0$

$$
\begin{aligned}
2yy' - 2xy' - 2y &= 0 \\
2yy' - 2xy' &= 2y \\
y'(2y - 2x) &= 2y \\
y' &= \frac{2y}{2y - 2x} = \frac{2y}{2(y - x)} \\
y' &= \left.\frac{y}{y - x}\right|_{(2,-3)} = \frac{-3}{-3 - 2} = \frac{-3}{-5} = \frac{3}{5}
\end{aligned}
$$

The correct choice is (E).

12. The average value of $\sqrt{3x}$ on $[0,9]$ is $\frac{1}{9-0}\int_0^9 \sqrt{3x}\,dx$.

$\int \sqrt{3x}\,dx = \sqrt{3}\int \sqrt{x}\,dx = \frac{2\sqrt{3}}{3}x^{\frac{3}{2}}$.

Therefore, the average value is $\frac{1}{9}\left(\frac{2\sqrt{3}}{3}x^{\frac{3}{2}}\right)\Big|_0^9 = 2\sqrt{3}$.

The correct choice is (B).

13. The derivative represents the instantaneous change in C with respect to x; thus its units are the units of C divided by the units of x: dollars per item. I is true.

The derivative represents the additional cost of producing one more item (i.e. the change in cost per item produced). II is true.

The units for the rate at which items are produced would be items per time. III is false.

The correct choice is (D).

14. If g is the inverse function of f, then $f(g(x)) = x$.

The expression $f'(g(0))\,g'(0)$ is the derivative of $f(g(x))$ at $x = 0$.

However, $f(g(x)) = x$ so its derivative is always 1.

The correct choice is (B).

15. $y = 6x^2 + \frac{x}{2} + 3 + 6x^{-1}$

$y' = 12x + \frac{1}{2} - 6x^{-2}$

$y'' = 12 + 12x^{-3}$

$y'' = 0$ when $12 + 12x^{-3} = 0$ or $12(1 + x^{-3}) = 0 \Rightarrow x = -1$

Also note that y'' does not exist at $x = 0$.

Signs of y'': + (left of −1), − (between −1 and 0), + (right of 0)

Therefore, the graph of y is concave down for $-1 < x < 0$, where $y'' < 0$.

The correct choice is (B).

16. Let (x, y) be a point on the parabola $y = 12 - x^2$.

Then the area of the rectangle, $A = (2x)(12 - x^2) = 24x - 2x^3$.

$A' = 24 - 6x^2 = -6(x^2 - 4) = -6(x-2)(x+2) \Rightarrow A' = 0$ when $x = \pm 2$.

The maximum occurs when $x = 2$. So the area of the rectangle is $A(2) = 32$.

The correct choice is (D).

17. The units of an accumulation function are the product of the units of the dependent and independent variables. Thus

$$\frac{liters}{kilometer} \cdot kilometers = liters$$

If the function $r(c)$ were graphed, the definite integral would give the area under the graph. The units of the horizontal dimension of the "squares" are those of the independent variable, kilometers. The units of the vertical dimension are those of the dependent variable, liters/kilometer. The area of the squares is the product namely liters.

The correct choice is (A).

18. Use the Product Rule to find $f'(x)$:

$$f'(x) = (e^x)\left(\frac{1}{x}\right) + (\ln x)(e^x)$$

$$= \frac{e^x}{x} + e^x \ln x$$

$$f'(e) = \frac{e^e}{e} + e^e \ln e$$

Since $\ln e = 1$, $f'(e) = \frac{e^e}{e} + e^e$ or $e^{e-1} + e^e$.

The correct choice is (A).

19. By the Fundamental Theorem of Calculus (version II),

$\frac{d}{dx}\int_0^{2x} (e^t + 2t)\,dt = 2(e^{2x} + 2(2x)) - 0 = 2e^{2x} + 8x.$

The correct choice is (E).

20. The area enclosed by the graph and the axis is the distance the object has moved from the starting point. When the velocity is positive (above the axis) the object is moving to the right; when the velocity is negative (below the axis) the object is moving to the left. The object moves to the right from the beginning past A and B to point C when the objects begins moving to the left. Therefore, the object is farthest to the right at C.

The correct choice is (C).

21. $f'(x) = 3x^2 \Rightarrow f(x) = x^3 + C$

Since $f(-1) = 2$, $2 = (-1)^3 + C$, or $C = 3$.

Therefore, $\int_0^2 (x^3 + 3)\,dx = \left.\frac{x^4}{4} + 3x\right|_0^2 = (4 + 6) - (0 + 0) = 10.$

The correct choice is (D).

22. The volume of a solid of a known cross section area $A(x)$ from $x = a$ to $x = b$ is $V = \int_a^b A(x)\,dx$.

Since $4x^2 + y^2 = 1$, we have $y = \pm\sqrt{1-4x^2}$ and the area of a semicircle is $(1/2)\,\pi y^2$.

(The ellipse $4x^2 + y^2 = 1$ has x-intercepts at $x = \pm 1/2$).

$$\text{The volume is } \int_{-\frac{1}{2}}^{\frac{1}{2}} (1/2)\,\pi y^2 dx = \int_{-\frac{1}{2}}^{\frac{1}{2}} (1/2)\,\pi(1-4x^2)\,dx = \pi\int_0^{\frac{1}{2}} (1-4x^2)\,dx$$

$$= \pi\left(x - \frac{4x^3}{3}\right)\Bigg|_0^{\frac{1}{2}}$$

$$= \frac{\pi}{3}$$

The correct choice is (C).

23. Since $\sin^2 x = 1 - \cos^2 x = (1-\cos x)(1+\cos x)$,

$$f(x) = \frac{\sin^2 x}{1-\cos x} = \frac{(1-\cos x)(1+\cos x)}{1-\cos x} = 1 + \cos x$$

Therefore, $f'(x) = -\sin x$.

The correct choice is (C).

24. $\frac{dy}{dx} = y\cos x \Rightarrow \frac{dy}{y} = \cos x\,dx$

$$\Rightarrow \int \frac{dy}{y} = \int \cos x\,dx$$

$$\Rightarrow \ln|y| = \sin x + C$$

$$\Rightarrow |y| = e^{\sin x + C} = e^C e^{\sin x}$$

$$\Rightarrow y = C_1 e^{\sin x}$$

Since $y = 3$ when $x = 0$, $3 = C_1 e^{\sin x}$

$$3 = C_1\,(1)$$

Therefore, $y = e^{\ln 3}\cdot e^{\sin x}$ or $y = 3e^{\sin x}$

The correct choice is (E).

25. The number of subintervals in this situation does not matter. To compare the relative sizes draw one rectangle or trapezoid on the interval $[a, b]$. The relative sizes of these rectangles will be true for any number of rectangles.

The left Riemann sum rectangle(s) will be larger than any of the others.

The correct choice is (B).

26. $f(x) = \ln(3x+2)^k$

$= k\ln(3x+2)$

$$f'(x) = k \cdot \frac{1}{3x+2} \cdot 3 = \frac{3k}{3x+2}$$

Since $f'(2) = 3$, $\frac{3k}{8} = 3 \Rightarrow 3k = 24$ or $k = 8$.

The correct choice is (D).

27. The various values of g are found by calculating the areas of the regions between the graph and the x-axis. Those regions below the axis are counted as negative values.

$g(-4) = -1.5 = -2 + 0.5$

$g(-3) = -2$

$g(-1) = 0$

$g(0) = 1.5 = 0.5(1+2)$

$g(3) = 0 = 2 - 2$

Hence $g(-1) = g(3)$.

The correct choice is (C).

28. Let $u = f(t) \Rightarrow du = f'(t)\,dt$

$$\int \frac{f'(t)\,dt}{f(t)} = \int \frac{du}{u} = \ln u$$

So, $\displaystyle\int_a^b \frac{f'(t)\,dt}{f(t)} = \ln[f(t)]\Big|_a^b = \ln[f(b)] - \ln[f(a)]$

Since $\displaystyle\int \frac{f'(t)\,dt}{f(t)} = \ln[f(b)], \ln[f(b)] - \ln[f(a) = \ln[f(b)]],$

and $\ln[f(a)] = 0 \Rightarrow f(a) = 1.$

The correct choice is (C).

29. Let $u = 2x$, so $du = 2\,dx \Rightarrow dx = \frac{1}{2}du$, when $x = 2, u = 4$ and when $x = 3, u = 6$.

$$\int_2^3 f(2x)\,dx = \frac{1}{2}\int_4^6 f(u)\,dx \Rightarrow \int_4^6 f(u)\,du = 2\int_2^3 f(x)\,dx$$

Therefore, $\displaystyle\int_4^6 f(x)\,dx = 2(8) = 16$

The correct choice is (D).

30. The average rate of change of $f(x)$ on $[a,b]$ is $\dfrac{f(b)-f(a)}{b-a}$.

In this problem, the average rate of change is $\dfrac{f(2)-f(0)}{2-0} = \dfrac{16-0}{2-0} = 8.$

Note: Do not confuse the average rate of change of f with the instantaneous rate of change, which is the derivative at a given point.

The correct choice is (B).

31. The graph of $f(x)$ has 2 relative maxima and 2 relative minima, which means that $f'(x)$ crosses the x-axis in 4 points.

The correct choice is (E).

32. Use implicit differentiation to find the derivative:

$$3x^2 + 6x + 2y\frac{dy}{dx} = 0$$
$$\frac{dy}{dx} = \frac{-3x^2 - 6x}{2y}$$

The derivative is undefined and the tangent is vertical when $y = 0$. When $y = 0$,

$$x^3 + 3x^2 = 4$$

Solve this equation on a calculator.

$$x = -2, 1$$

So the points are $(-2,0)$ and $(1,0)$.

The correct choice is (D).

33. The profit function $P(x) = R(x) - C(x)$, where

$R(x)$: revenue (sales) function

$C(x)$: cost function

If x calculators are sold and the price per calculator is $75 - 0.01x$ dollars each,

then $R(x) = x(75 - 0.01x)$ and $C(x) = 1850 + 28x - x^2 + 0.001x^3$

Thus the profit function,

$$P(x) = x(75 - 0.01x) - (1850 + 28x - x^2 + 0.001x^3)$$
$$= -0.001x^3 + .99x^2 + 47x - 1850$$

To maximize $P(x)$, find $P'(x) = -0.003x^2 + 1.98x + 47$

Solve this equation on a calculator. Thus $x = 683$.

Note: Since $P''(x) = -0.006x + 1.98$ and $P''(683) < 0 \Rightarrow x = 683$ is a maximum.

The correct choice is (C).

34. The area of the shaded region is bounded by the x-axis and $y = f(x)$.

Since $y = 0$ lies above the curve $y = f(x)$on $[a,b]$, the area is

$$\int_a^b [0 - f(x)]\, dx = \int_a^b -f(x)\, dx = -\int_a^b -f(x)\, dx \text{ or } \int_b^a f(x)\, dx.$$

Keep in mind, that $\int_a^b f(x)\, dx = -\int_b^a f(x)\, dx$.

The correct choice is (B).

35. Refering to the accompanying diagram,

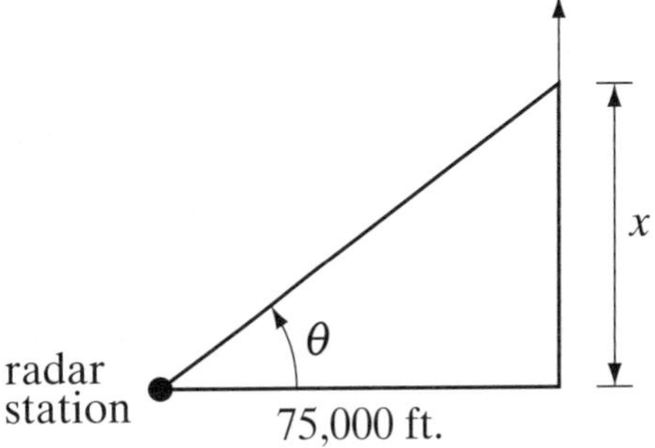

$$\tan\theta = \frac{x}{75{,}000} \Rightarrow \theta = \tan^{-1}\left(\frac{x}{75{,}000}\right).$$

Therefore, $\dfrac{d\theta}{dt} = \dfrac{(1/75{,}000)(dx/dt)}{1 + (x/75{,}000)^2}$

Since $\dfrac{dx}{dt} = 16{,}500$ when $x = 38{,}000$

$$\frac{d\theta}{dt} = \frac{16{,}500/75{,}000}{1 + (38{,}000/75{,}000)^2} = \frac{165/750}{1 + (38/75)^2} \approx 0.175$$

The correct choice is (A).

36. Since the function is continuous, the limit at $x = 2$ must equal the value at $x = 2$. The value, determinted from the first part of the definition, is 3; so I is true. The continuity of f also implies that in the second part of the definition, $a(2) + 4 = 3$ and therefore $a = -\frac{1}{2}$. Hence the slopes of the two linear pieces are not equal and there is no derivative at $x = 2$; II is false. Since the two pieces are linear, there is no concavity, so there can be no place where the concavity changes; III is false. (Alternatively, the second derivative is zero everywhere (except at $x = 2$) and this also indicates there is no change in concavity.)

The correct choice is (A).

37. On the graph of $y = e^x$, y will equal 2 when $x = \ln 2$.

Therefore, the volume is $V = \pi\int_0^{\ln 2} [(2)^2 - (e^x)^2]\, dx = \pi\int_0^{\ln 2} (4 - e^{2x})\, dx$

The correct choice is (D).

38. The graph of a function will be concave up when $f''(x) > 0$. Since $f''(x) > 0$ for $x < 0$ and $b < x < c$, therefore I is true.

The (relative) maximum and minimum values of a function cannot be determined from the second derivative alone, therefore II is false.

The function f has points of inflection where $f''(x)$ changes sign. This occurs at $x = 0$ and $x = b$. So $x = 0$ and $x = b$ are points of inflection, therefore III is true.

The correct choice is (D).

39. Since $y = 2x - 3$ is the tangent line to f at $x = 1$, $(1, -1)$ is on the graph of f and $f'(1) = 2$, since the line $y = 2x - 3$ has a slope of 2.

Also since $g(x) = \frac{f(x)}{x}$, $g(1) = \frac{f(1)}{1} = -1$.

$g'(1)$ is obtained by finding $g'(x)$ using the Quotient Rule:

$$g'(x) = \frac{x\,f'(x) - f(x)}{x^2} \Rightarrow g'(1) = \frac{(1)\,f'(1) - f(1)}{1^2} = \frac{2-(-1)}{1} = 3$$

Since $g'(1) = 3$ and $g(1) = -1$, the equation of the line tangent to the graph of g at $x = 1$ is $y - (-1) = 3(x - 1)$ or $y + 1 = 3x - 3$, or $y = 3x - 4$.

The correct choice is (A).

40. The relative maximum values of a function occur where the first derivative of the function changes from positive to negative. By the Fundamental Theorem of Calculus the derivative is

$$f'(x) = \sin\left(\frac{1}{x}\right)$$

Graph the derivative in a convenient window such as x [0.1, 0.4] by y [-1.3, 1.3]

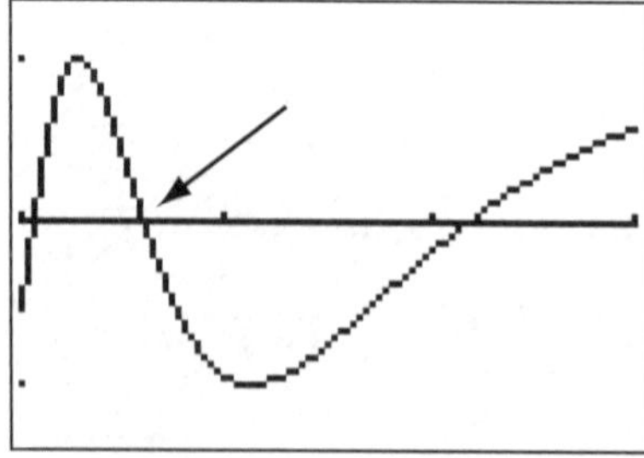

Using a calculator find the x-coordinate of where the derivative changes from positive to negative (indicated by the arrow).

The value is $x = 0.159$.

The other two values, $x = 0.106$ and $x = 0.318$, are the locations of the relative minimums of the function.

The correct choice is (B).

41. The definition of the derivative of f at $x = 2$ is: $f'(2) = \lim_{x \to 2} \frac{f(x) - f(2)}{x - 2}$.

Therefore, from the given information, $\lim_{x \to 2} \frac{f(x)}{x - 2} = \lim_{x \to 2} \frac{f(x) - f(2)}{x - 2} \Rightarrow f(2) = 0$, so I is true.

Since $f'(2) = 0$ and $f(2) = 0$, the graph of $f(x)$ has a horizontal tangent at the point (2,0), so III is true. Note: $f(x)$ will be tangent to the x-axis at $x = 2$.

If a function is differentiable at a point, it must be continuous there. Therefore, since f is differentiable at $x = 2\,(f'(2) = 0)$, f is continuous at $x = 2$, and II is true.

The correct choice is (E).

42. The mass of a piece of the rod of length Δx meters is found by multiplying $\rho(x)$ by Δx. The mass of these pieces are added giving $\sum_{i=1}^{n} \rho(x_i)\,\Delta x$. The limit of this Riemann sum as $n \to \infty$ is given by $\int_0^1 \rho(x)\,dx = \int_0^1 (1 + (1 - x)^2)\,dx = \frac{4}{3} = 1.333$

The units are $\frac{\text{grams}}{\text{meters}} \cdot \text{meters} = \text{grams}$

The correct choice is (B).

43. To write the equation of the tangent line, first find the slope:

$$\begin{aligned} f'(x) &= 3x^2 - 12x + 7 \\ f'(4) &= 3(4^2) - 12(4) + 7 \\ f'(4) &= 7 \\ f(4) &= -1 \\ y &= 7(x - 4) + (-1) \\ y &= 7x - 29 \end{aligned}$$

Then $f(4.2) = 0.648$ and $y(4.2) = 0.4$. The error is $|0.648 - 0.4| = 0.248$

The correct choice is (C).

44. $\int_0^{1000} 8^x\,dx - \int_a^{1000} 8^x\,dx = \int_0^{1000} 8^x\,dx + \int_{1000}^{a} 8^x\,dx = \int_0^{a} 8^x\,dx = \left.\frac{8^x}{\ln 8}\right|_0^a = \frac{8^a - 1}{\ln 8}$

Solving $\frac{8^a - 1}{\ln 8} = 10.40$ on a graphing calculator results in $a = 1.5$.

The correct choice is (B).

45. The average speed is given by $\frac{1}{32{,}000}\int_0^{32{,}000} s(a)\,da$. While it is possible to write a piecewise function for $s(a)$ it is easier to find the value of the integral by find the area under the graph directly.

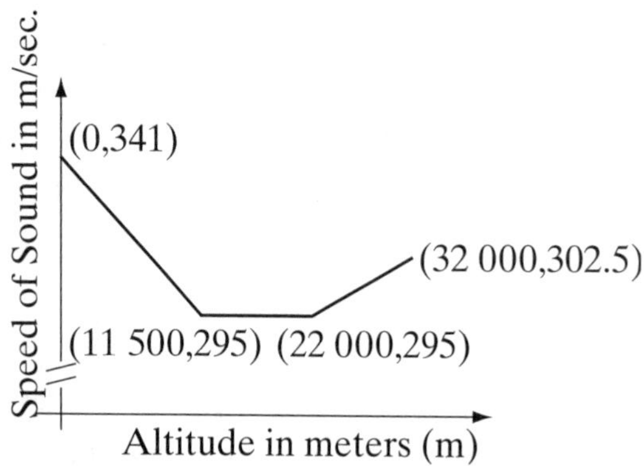

Region A is a trapezoid. The area is $\frac{1}{2}(11{,}500)(341 + 295) = 3{,}657{,}000$
Region B is a rectangle. The area is $295\,(10{,}5000) = 3{,}097{,}500$
Region C is a trapezoid. The area is $\frac{1}{2}(10{,}000)(302.5 + 295) = 2{,}987{,}500$
The total area is 9,742,000
Divide by 32,000 to find the average value = 304.4 m/sec.

Note: It is incorrect to find the average speed on each part and then average the averages. This gives answer (B) 303.9.

The correct choice is (C).

1a.

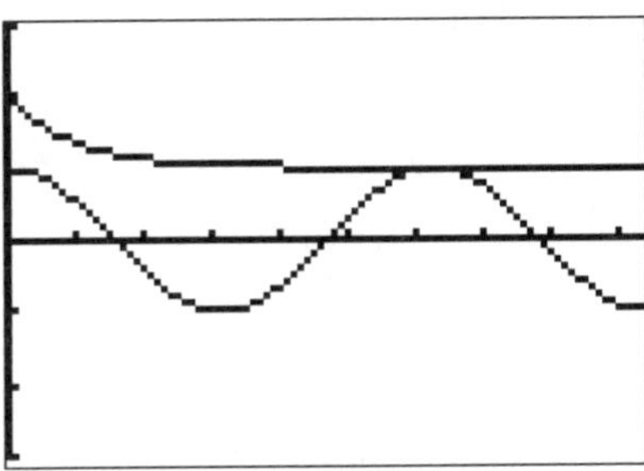

1b. Since for all x, $1 + e^{-x} > 1$ and $\cos x \leq 1$, the particles can never be at a point with the same y-coordinate, therefore they will never collide.

1c. The distance between the particles is $s(x) = 1 + e^{-x} - \cos x$. Differentiate to find the critical values:

$$\frac{ds}{dt} = -e^{(-x)} + \sin x$$

Set the derivative equal to zero and solve for x on a calculator. $x = 0.5885$, $x = 3.0963$, $x = 6.2850$, and $x = 9.4246$

Then $s(0.5885) = 0.7233$; $s(3.0963) = 2.04419$; $s(6.2850) = 0.00186$; $s(9.4246) = 2.0000806$

Now check the endpoints $s(0) = 1$ and $s(3\pi) = 2.0000807$

The minimum distance is about 0.00186 or 0.001 or 0.002

1d. From part (c) the maximum distance is 2.04419 or 2.044

2a. Separate the variables, integrate and solve for R:

$$\frac{dR}{R} = k \cdot dt$$

$$\ln R = k \cdot t + C_1$$

$$R = C \cdot e^{kt}$$

Substitute $t = 0$ and $R = 10^6$ and find C: $10^6 = Ce^0$ and $C = 10^6$

Substitute $t = 100$ and $R = 10^2$ to find k:

$$10^2 = 10^6 e^{100k}$$

$$10^{-4} = e^{100k}$$

$\ln 10^{-4} = 100k$ and $k = \dfrac{\ln 10^{-4}}{100} \approx -0.092$

Therefore, $R(t) = 10^6 e^{-0.092t}$

2b. Solve $10 = 10^6 e^{-0.092t}$ for t on a calculator or by hand:

$$10^{-5} = e^{-0.092t}$$

$$\ln 10^{-5} = -0.092t$$

$t = \dfrac{\ln 10^{-5}}{-0.092} = 125.140$ or about 125 seconds

2c. Solve $\frac{1}{2}10^6 = 10^6 e^{-0.092t}$

$$\tfrac{1}{2} = e^{-0.92t}$$

$$\ln \tfrac{1}{2} = -0.092t$$

$t = \dfrac{\ln \frac{1}{2}}{-0.092} = 7.534$ seconds

<u>REMINDER</u>: Numerical and algebraic answers need not be simplified, and if simplified incorrectly credit will be lost.

3a. The acceleration is negative when the velocity is decreasing. This occurs during the interval $15 \leq t \leq 45$.

3b. The average acceleration is the change in the velocity divided by the change in time:

$$\frac{3.1 - 0}{15 - 0} = 0.207 \text{ feet/second/second}$$

3c. Using a right Riemann sum, a left Riemann sum or trapezoidal approximation, all give the same value. The left hand sum is

$$\frac{30 - 0}{6}(0 + 1.6 + 2.7 + 3.1 + 2.7 + 1.6) = \frac{30}{6}(11.7) = 58.5$$

3d. The integral evaluated in part (c) is an approximation of how high up the rider traveled in the first half of one complete revolution of the ferris wheel, this is the diameter. The ferris wheel is approximately 58.5 feet in diameter.

4a. Area $\int_0^1 (e^x - e^{-x})\,dx = e^x + e^{-x}\Big|_0^1$

$$= e + e^{-1} - (e^0 + e^0)$$

$$= e + e^{-1} - 2$$

4b. Volume $= \int_0^1 \pi((e^x)^2 - \pi(e^{-x})^2 dx) = \pi\left(\frac{e^{2x}}{2} - \frac{e^{-2x}}{-2}\right)\Big|_0^1$

$$= \pi\left(\frac{e^2}{2} + \frac{e^{-2}}{2} - \left(\frac{1}{2} + \frac{1}{2}\right)\right) = \pi\left(\frac{e^2 + e^{-2} - 2}{2}\right)$$

4c. The length of the base of each rectangle is $(e^x - e^{-x})$ and the cross-section area is $x(e^x - e^{-x})$.

Thus the Volume $= \int_0^1 x(e^x - e^{-x})\,dx$

5a. When $x = -1$ the point $(-1, 6a)$ lies on $f(x)$; at this point the slope is found from the derivative $f'(x) = -2ax$; $f'(-1) = -2a(-1) = 2a$. The equation of the tangent line may be expressed as:

$$y - (6a) = (2a)(x - (-1)) \text{ or } y = 2ax + 8a$$

5b. Substitute $x = 0$ into the general equation of the tangent line, $y = 2a(0) + 8a$. The y-intercept is the point $(0, 8a)$.

5c. Substituting $y = 0$ into the equation of the tangent line, and solving for x: $0 = 2ax + 8a$ and $x = -4$. The x-intercept is $(-4, 0)$ and does not depend on a.

5d. The area is given by

$$\int_{-1}^{0} [(2ax + 8a) - a(7 - x^2)]\, dx = ax^2 + 8ax - 7ax + \frac{ax^3}{3}\Big|_{-1}^{0} = 0 - \left(a - 8a + 7a - \frac{a}{3}\right) = \frac{a}{3}$$

6a. As t increases $e^{-0.02t}$ gets smaller. Therefore the function approaches 1 from above. The limit is 1.

6b. $f'(t) = -0.02\,(0.9)\,e^{-0.02t}$. Since the derivative is always negative, the function is always decreasing.

6c. $f(0) = 1 + 0.9e^{0} = 1.9$. The function's values start at 1.9 and decrease towards 1. Thus the range is $1 < f(x) \leq 1.9$ or $(1,1.9]$.

6d. The function is decreasing most rapidly when the derivative has its smallest value. The second derivative of the function is $f''(t) = (-0.02)^{2}\,(0.9)\,e^{-0.02t}$. The second derivative is always positive indicating that the function is always concave up. This means that the slope, which is negative from part (b) and increasing towards zero, has its smallest value at the left end of the domain, $t = 0$. At $t = 0$ is where the slope is decreasing most rapidly.